IMAGES
of Aviation

AVIATION IN READING AND BERKS COUNTY

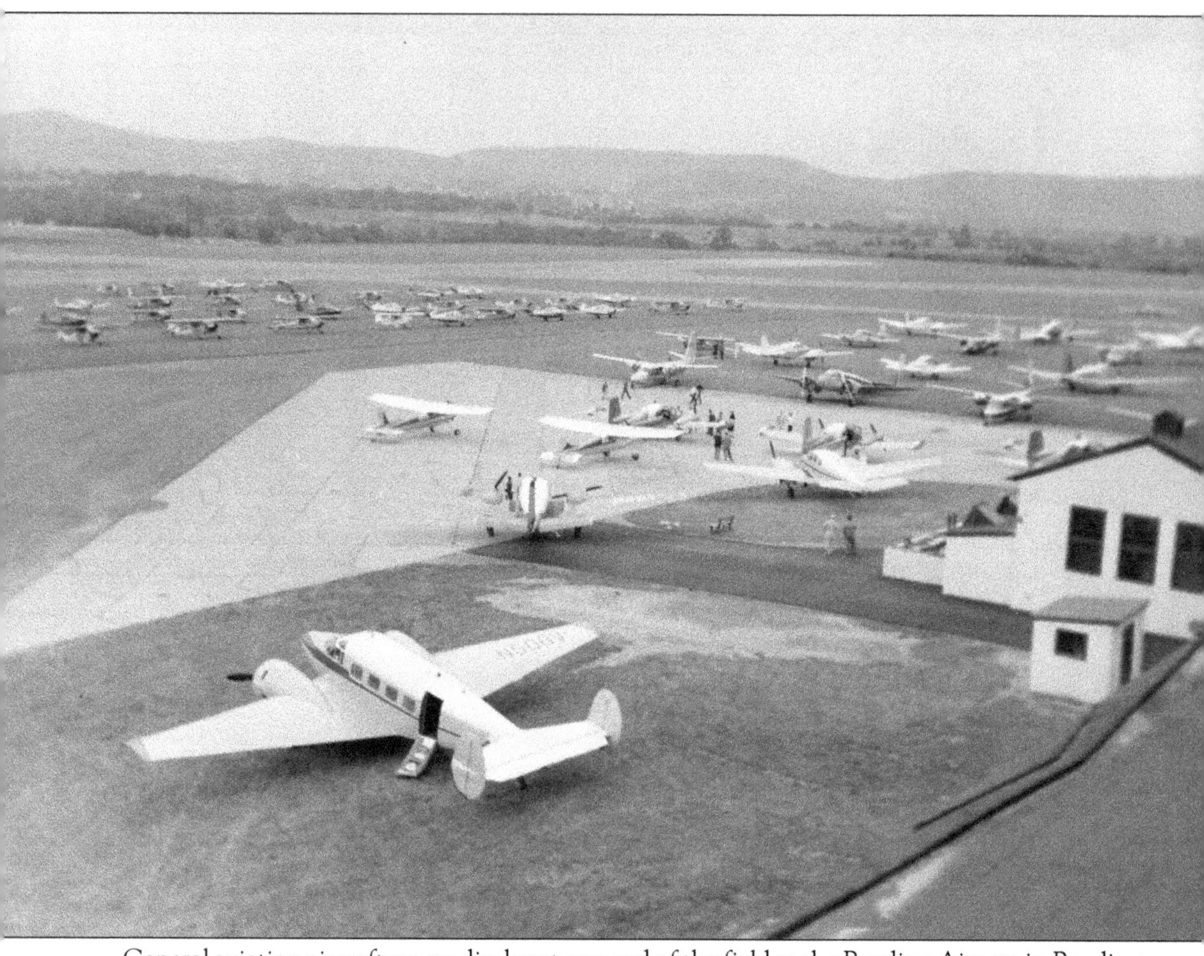

General aviation aircraft are on display at one end of the field at the Reading Airport in Reading, Pennsylvania, during the National Maintenance and Operations Meeting on June 4, 1955. Years later, the event would be renamed the Reading Airshow, the second largest air show in the world until the late 1970s. (Courtesy of the National Air and Space Museum.)

On the Cover: Eddie Nibur (center, with pipe) is with his operations team at Madeira Field in front of an 11-passenger Stinson Tri-Motor Deluxe for the airfield's air carnival on June 22, 1935. From left to right are Quinton Cudney, Joe Levan, Nibur, Hugh Sullivan, and Charlie Douglass. (Courtesy of Paul Nibur.)

IMAGES
of Aviation

AVIATION IN READING AND BERKS COUNTY

Michael J. Floriani

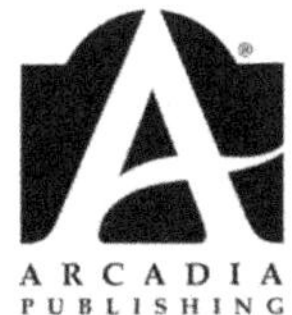

ISBN 978-1-4671-0322-0

Published by Arcadia Publishing
Charleston, South Carolina

Library of Congress Control Number: 2018962856

For all general information, please contact Arcadia Publishing:
Telephone 843-853-2070
Fax 843-853-0044
E-mail sales@arcadiapublishing.com
For customer service and orders:
Toll-Free 1-888-313-2665

Visit us on the Internet at www.arcadiapublishing.com

To my late father, Frank D. Floriani, who gave me countless memories of Reading aviation. His love of aviation became mine—and with it came this book.

CONTENTS

Acknowledgments 6

Introduction 7

1. Before Powered Flight 9
2. Barnstorming and the Early Days 15
3. Berks County Pioneer Aviators 23
4. Airfields and Aerial Navigation 43
5. Air Shows and Events 75
6. Military Aviation 93
7. Commercial and General Aviation 107
8. Aviation-Related Industries 119

ACKNOWLEDGMENTS

My greatest experience in creating this book was researching individuals who had a significant impact on Berks County aviation. In most cases, those who did play a role are no longer with us; however, it afforded me the opportunity to communicate with relatives or friends of those aviators. One person in particular is Paul Nibur, son of the late Eddie Nibur, an excellent aviator and extraordinary businessman with a strong work ethic and vision of aviation that many did not possess at the time. Paul was not only willing to respond to my numerous questions about his father's experiences, but also willing to share his father's personal autobiography with me. Equally important to our county's aviation history was businessman and aviator Walter Grimes and aviatrix Mildred Zimmerman. John Grimes, his grandson, was generous to share and entrust me with some very rare aviation pictures of Grimes Airport, while Cathryn Appel and her husband, Jim, were especially accommodating to my last-minute requests for information about Cathryn's inspiring mother.

Historical pictures and information about Reading Airport would not have been possible without the assistance of Terry Sroka, Reading Airport manager. I cannot express my gratitude enough for helping me when, I am sure, more pressing demands were upon him for the daily operations of the airport. Also, my appreciation goes to those within the Facebook community that were able to share their pictures and aviation anecdotes with me. To those members allowing me to use these images, I have given credit where appropriate.

Charles J. Adams III and George M. Meiser IX, local historians and authors, made me aware of certain aviation pioneers or answered my questions about historical events. Also, I wish to thank Sime Bertolet, executive director of the Berks History Center, for sharing pictures of his late father, founder of Reading Aviation Service, Suburban Airlines, and the Reading Airshow.

Finally, I wish to thank my wife, family, and friends who encouraged me and helped me with this project when years passed since I initially started it. Without them, this book would still be a dream.

INTRODUCTION

As an aviator, I wondered what the region looked like in the early days of aviation. My purchase of a 1946 aviation sectional chart 13 years ago led me to think about the airfields that populated that chart and the stories that were lost when those airfields were abandoned and converted into retail shopping centers or housing developments. Brave aviators flew on those fields while some earlier aviators even found their own fields suitable for takeoffs and landings. The rich stories that encompassed those years are mostly forgotten; however, I hope I have visually recaptured some of the simplicity of those days. Moreover, I hope I have made a faithful effort to honor those who played a role in the rich heritage of aviation in Reading and throughout Berks County.

Berks County aviation began very much like it did in other parts of America, with aeronauts performing balloon ascensions in the center of Reading. It did not take long for residents to become bored by those events, forcing these early aviators to perform death-defying parachute leaps and aerial acrobatics. The same held true some years later as barnstormers performed aerial stunts to attract local residents to various fields suitable for landings.

Well before the beginning of powered flight, the citizens of Reading and Berks County probably didn't give much credence to an article published in the *Reading Times* on January 14, 1898—almost six years before Wilbur and Orville Wright made their first flight—discussing the possibility of powered flight. Since then, man has overcome many of the challenges of flight and made significant advances in aviation. In Reading and Berks County, aviation began with individuals willing to take chances in life (and with their lives) to soar above the rolling hills, towns, and farmlands in their balloons, gliders, and airplanes. Some individuals attempted to build their own aircraft, while others came from different regions across the United States to entertain the locals with their flying skills well before the government began to regulate aviation, which ended barnstorming by 1929.

Almost six years after the Wright brothers achieved their first successful flight at Kitty Hawk, North Carolina, local newspapers wrote about a native German living in a barn in Alsace Township. The inventor, fearful of disclosing his location, was unwilling to share much information on solving the problem of human flight. His airship, much larger and advanced in design, was being prepared for the 1909 Reading fair. Unfortunately, neither my research nor history proved whether his aircraft ever achieved the goal that Frank W. Boyd set for himself.

In August 1912, on a field at the former Carsonia Park, Berks Countians witnessed the county's first flight. Charles F. Walsh, an aviator from San Diego, was contracted to perform several aeroplane flights as part of a Reading-based bakery promotional event. These successful flights made Walsh an instant hero with the day's crowd, which exceeded 10,000. Tragically, Walsh died exactly two months later in Trenton, New Jersey, from a structural failure of his aircraft at 2,000 feet.

The 1920s ushered in the greatest increase in barnstorming activity before the Great Depression made many operators bankrupt. During that decade, Reading's first airport was formed to manage Reading Airways, the first company in Pennsylvania to receive a charter for commercial flying.

Whander Field, Reading's first commercial airport, was dedicated on August 16, 1927, but other older airfields existed in the county.

Although formally dedicated in 1934, Madeira Field began in 1919 when Charles Madeira, a farmer living in Hyde Park, allowed pilots to use his field for aerial circuses and passenger hopping. Once known as the oldest Berks airport in operation, Madeira field became a significant contributor to the county's aviation community and economy when Eddie Nibur established flight instruction and charter operations on the field. Nibur grew the airport into a first-class operation until he left in the 1940s to fly for United Airlines. The business was sold to Raphael Rentschler until the airfield was closed in 1963 and developed into today's Madeira Shopping Center. Years later, Nibur gained national notoriety when in 1969, he faced the barrel of a hijacker's gun and was forced to fly from Los Angeles to Cuba—the longest hijack in United Airline's history at the time.

Berks County has been privileged to have other aviation pioneers contribute to the industry in various ways. Cliff Hadley became one of the first airmail pilots in the United States, J. Earl Steinhauer made manufacturing contributions while at Fairchild Aircraft, Sime Bertolet and R. Harding Breithaupt built a commercial aviation empire in Reading, and Carl Spaatz became the leading US combat air commander in World War II. Other pioneers were less fortunate to establish their local or national status in aviation history. Nineteen-year-old Betsy Ross of Stouchsburg attempted to achieve a high-altitude flight record of 20,000 feet in a Taylorcraft but fell short of her goal by 1,800 feet, dashed by faulty equipment installed in her airplane by the National Aeronautics Association. Nine years later, a similar incident occurred for Reading's aviatrix and 27-year-old mother Mildred Zimmerman. She, however, accomplished the world record one day following her failed flight by reaching 26,200 feet in her Taylorcraft.

By the late 1930s, Reading built a first-class airport that in the 1940s became a training site for the Civilian Pilot Training Program and an Army Air Force facility that housed German POWs in onsite barracks. After World War II, and beginning with the Korean War, the Pennsylvania Air National Guard operated from the airport.

As the 1950s ended, commercial aviation grew as wartime pilots came home and took jobs with the commercial airlines. Also, general aviation grew as many of the former wartime pilots chose recreational flying or took flying jobs for industries inside and outside of Berks County. Sime Bertolet and R. Harding Breithaupt leveraged this potential and built Reading Aviation Service (RAS), a commercial airline, charter operation, and aircraft sales and service business. Eventually, RAS became the impetus for Reading's greatest annual event—the Reading Airshow.

The Reading Airshow, which began modestly in 1939, eventually became the second-largest air show in the world as the annual event moved into the 1960s and 1970s. It was an economic boon to the area as hotels, restaurants, and vendors saw an influx of 100,000 daily attendees and over 600 airplanes.

From the late 1970s until today, aviation has realized significant changes and challenges resulting from increased regulation, technological developments, and pilot shortages. With challenges come opportunities to contribute to today's aviation industry for Berks County's current residents and businesses.

As the Reading Airport celebrates its 80th anniversary, this book commemorates not only the contributions the airport made to the county, but also honors those who were influential in establishing Reading and Berks County's aviation heritage. I hope that today's generation contributes to aviation like those visionaries in aviation's earlier days. The following images within this book are for all generations—to inspire younger generations while bringing back wonderful memories for older generations who experienced what aviation was like in those earlier days.

One

BEFORE POWERED FLIGHT

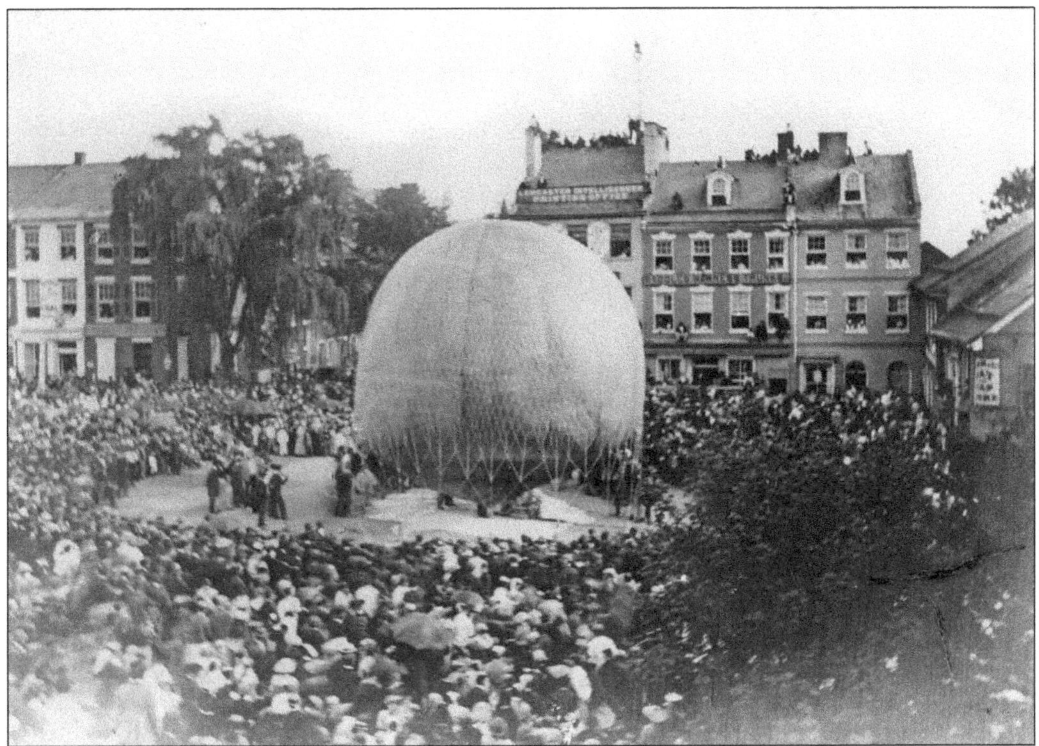

Although ballooning was not new by the 1800s, the first balloon ascension in Reading, similar to this one photographed from Lancaster's Penn Square, was conducted in August 1836 by John Wise of Lancaster. The flight occurred on Fifth Street between Franklin and Chestnut Streets. The balloon, inflated with hydrogen gas, experienced problems that shortened the flight to four or five miles. Wise chronicled these flights in his book *Through the Air*. (Courtesy of the Lancaster Historical Society.)

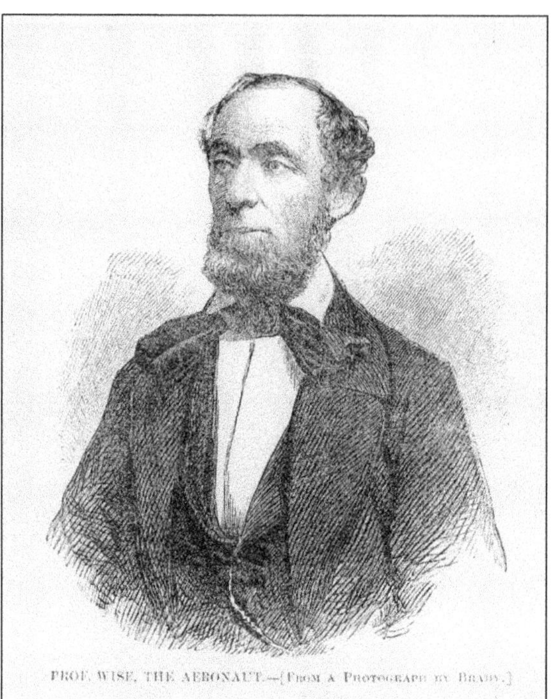

John Wise, born in 1808, became interested in ballooning when he read a newspaper article and began conducting aerial experiments tying his cat to a homemade parachute, observing its safe descent from a local church steeple. In 1835, he constructed a balloon and made his first ascent in Philadelphia. He described himself as "the world-renowned astronaut who had more voyages through the Heavens than any other man," completing 463 ascensions. (Courtesy of the Lancaster Historical Society.)

THROUGH THE AIR:

A NARRATIVE OF

FORTY YEARS' EXPERIENCE AS AN AËRONAUT.

COMPRISING

A HISTORY OF THE VARIOUS ATTEMPTS IN THE ART OF FLYING BY ARTIFICIAL MEANS FROM THE EARLIEST PERIOD DOWN TO THE PRESENT TIME.

WITH AN ACCOUNT OF

The Author's Most Important Air-Voyages

AND

HIS MANY THRILLING ADVENTURES AND HAIRBREADTH ESCAPES.

ALSO,

AN APPENDIX, IN WHICH ARE GIVEN FULL INSTRUCTIONS FOR THE MANUFACTURE AND MANAGEMENT OF BALLOONS.

BY JOHN WISE.

"Stand still, and consider the wondrous works of God.
"Dost thou know when God disposed them, and caused the light of his cloud to shine?
"Dost thou know the balancings of the clouds, the wondrous works of Him which is perfect in knowledge?"—JOB xxxvii.

Profusely Illustrated.

TO-DAY PRINTING AND PUBLISHING COMPANY,
PHILADELPHIA, NEW YORK, BOSTON &
AND CHICAGO.
1873.

Descriptions of Wise's ascensions and experiments are found in his 1873 book *Through the Air.* Overflying Reading, he described the city as "a handsome aspect: the white streets crossing at angles and beautiful spires and domes white as snow with glittering balls and vanes highly interesting." During the Mexican War, Wise devised a plan to take over a Mexican city by fabricating a balloon capable of dropping explosives on it. (Courtesy of the Lancaster Historical Society.)

Washington "Wash" Harrison Donaldson was a prodigious acrobat, performing dances and gyrations on rope to the amazement of Philadelphians. He became a resident of Reading through the persuasion of local amusement manager John D. Mishler. Donaldson lived on Sixth and Penn Streets and was considered the most daring American and greatest balloonist of his time. P.T. Barnum called him "the most daring and colorful personality ever known." (Author's collection.)

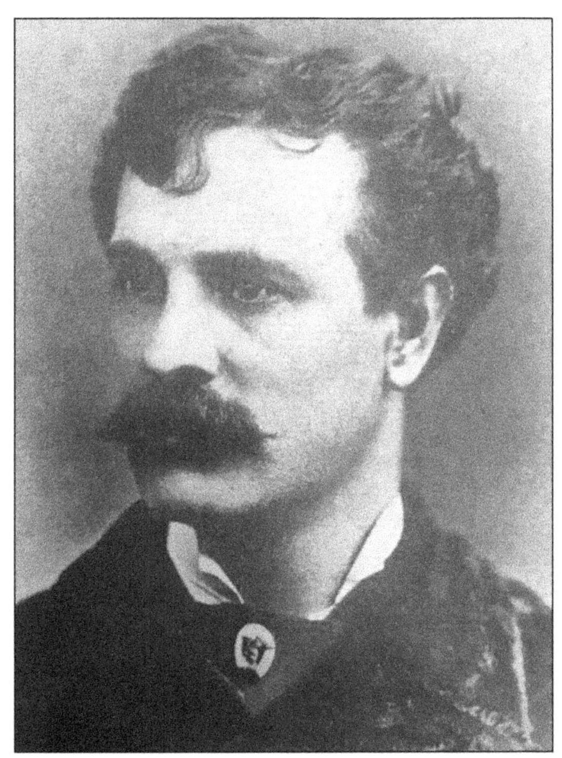

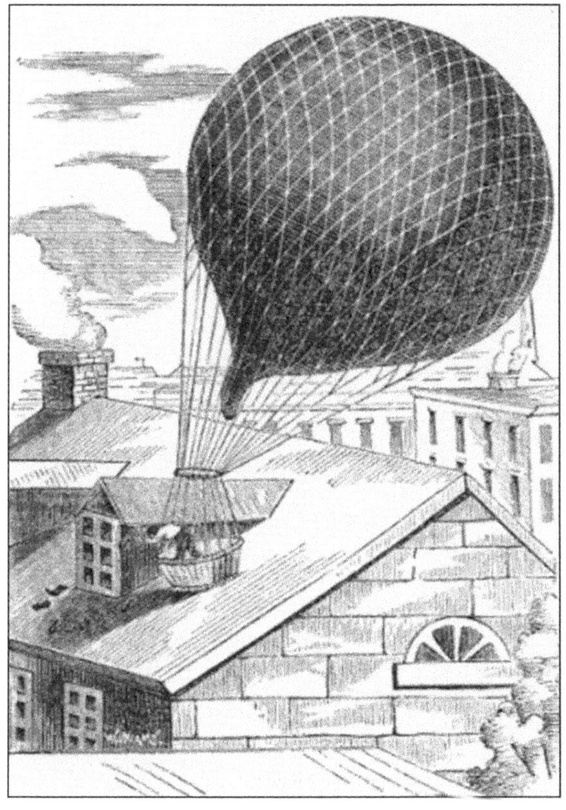

In 1871, Washington Donaldson prepared for his maiden ascension in a balloon named *Comet*. The flight began on Penn Street, where the balloon rose over homes and Donaldson performed gymnastic feats on a hoop which, according to the *Reading Eagle*, "made one's blood run cold." Another balloon ascension was made until the trapeze struck a roof, creating a hole in the balloon and causing a rapid descent. Donaldson was uninjured. (Author's collection.)

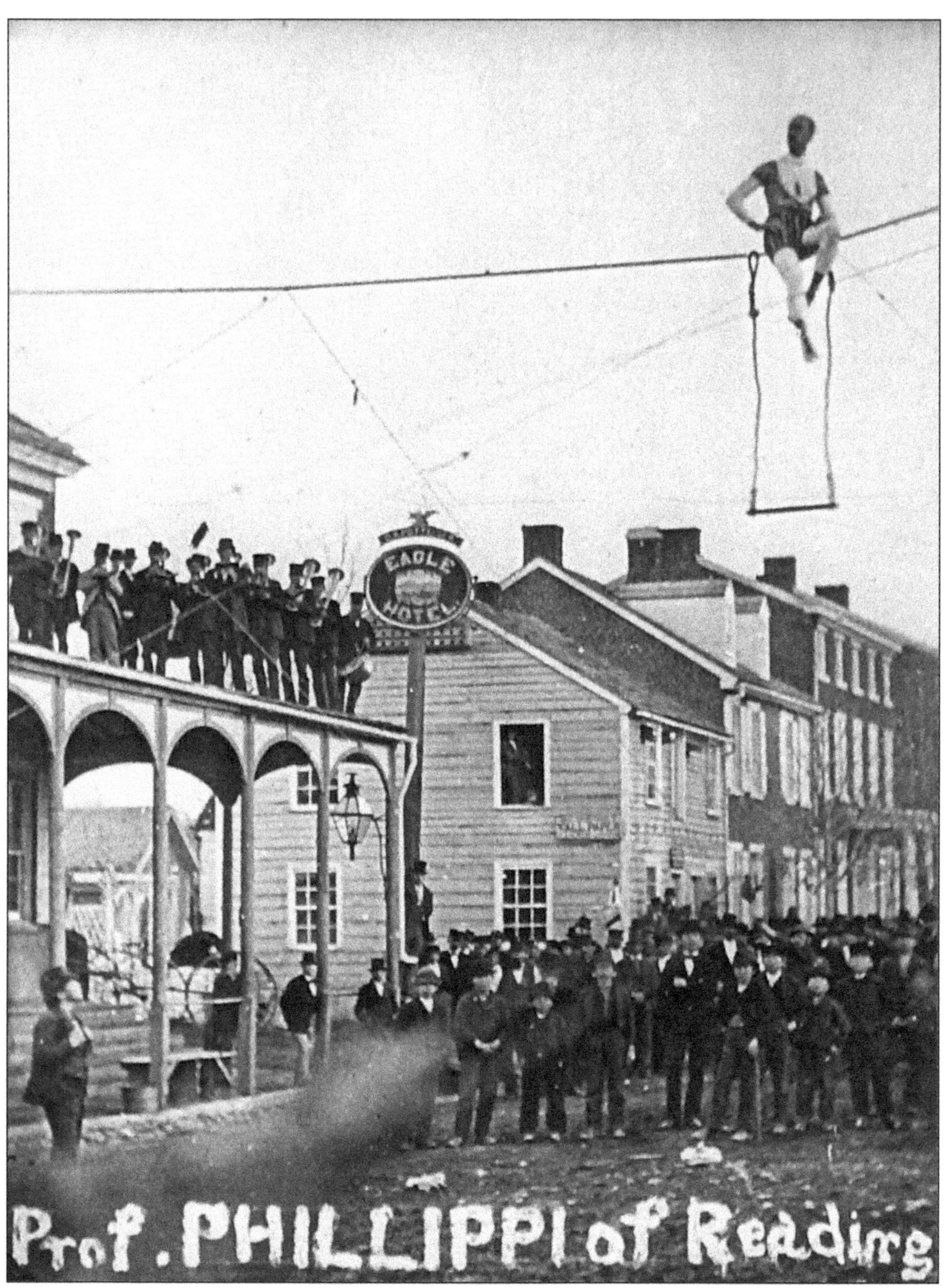

Jacob "J.R." Phillippi, a resident of 714 Cherry Street in Reading, was well known for performing ascensions throughout the country. At 17, he became interested in ballooning, assisting Washington Donaldson in his first ascension from Reading's Penn Square. Two years later, he held his first flight on Penn Square. On July 9, 1908, Phillippi was observed launching from Tenth and Court Streets in a balloon composed of 400 sheets of tissue paper. At 19 years old, Phillippi purchased a used balloon without a basket so that he could emulate Donaldson's tricks from a trapeze; however, his friends contacted Reading's mayor to ban the show. With Phillippi's assurance to the mayor that he would not attempt such tricks, he reneged on his word. After his flight, he received a hero's welcome. (Courtesy of Charlie Adams III.)

GRAND BALLOON ASCENSION!

Prof. JOHN H. STEINER,

The Champion American Aeronaut respectfully announces to the citizens of Reading and Berks county, that he will make his 631 GRAND ASCENSION.

From *CENTRE SQUARE*, in the City of Reading in his mammoth Balloon "Europa,"

On Saturday, June 11th, 1859.

Topical, or Partial Ascensions of from 300 to 500 feet will commence at 10 o'clock for the accommodation of all who desire to make the acquaintance of the upper regions. The final ascension will take place between 4 and 5 o'clock.

Tickets for Topical Ascension $2; and for the final or Grand Ascension 25 cts. per lb. for the weight of the passenger.

june 10 2t

A native Bavarian and resident of New York in 1853, John H. Steiner was known for his daredevil feats. He went to Reading on June 11, 1859, to perform his grand ascension at Penn Square. The Reading newspaper announced Steiner would make his 631st grand ascension from Centre Square in his mammoth balloon *Europa*, taking passengers to 300 to 500 feet. (Author's collection.)

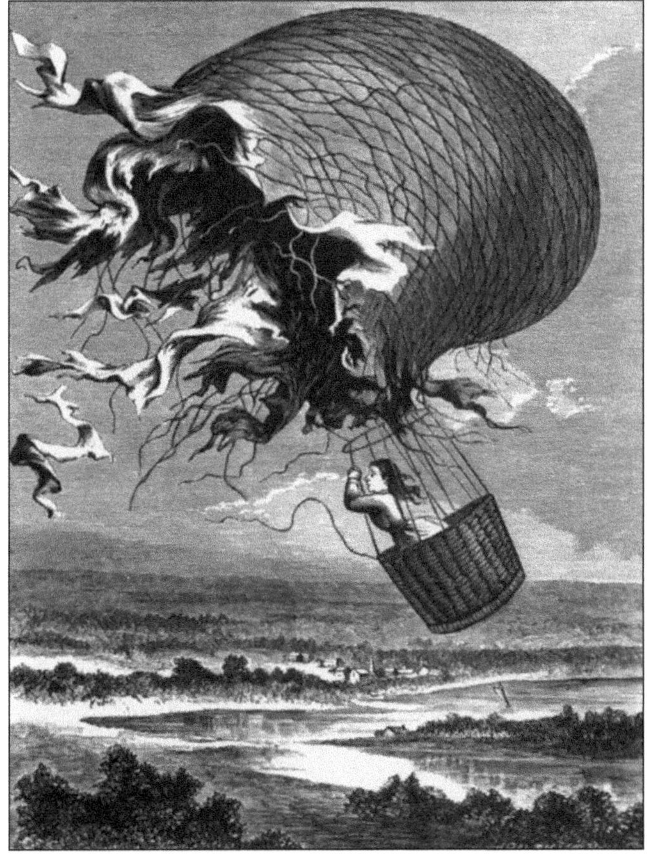

Considered the first American woman to fly alone in a gas-filled balloon, Louisa Bradley, who lived in Reading for a period, purchased a balloon from John Wise in October 1854 for $100 and received some instructions from the balloonist. The balloon, made of silk, was dry and cracked, resulting in a burst balloon. Fortunately, the torn silk acted as a parachute, which saved Bradley from injury or death. (Author's collection.)

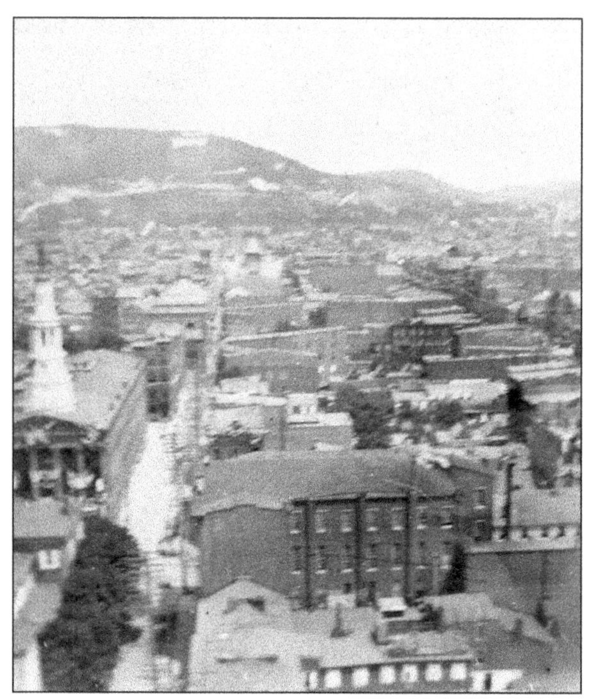

The first known aerial photographs of Reading came from a camera attached to a kite flown by William Abner Eddy of New York to celebrate Reading's sesquicentennial on June 5–12, 1898. Eddy used a specially rigged kite, developed through years of research by connecting multiple kites in tandem, to take these pictures with a Kodak camera looking east toward Mt. Penn (at left) and south at Fifth and Penn Streets (below). The kite reached an altitude of approximately 500 feet. Years later, others set a record altitude of 23,385 feet for pictures taken by a kite designed like that made by Abner Eddy in the late 1800s. His first pictures ever taken by kite were on May 30, 1895. (Both, courtesy of International Center of Photography.)

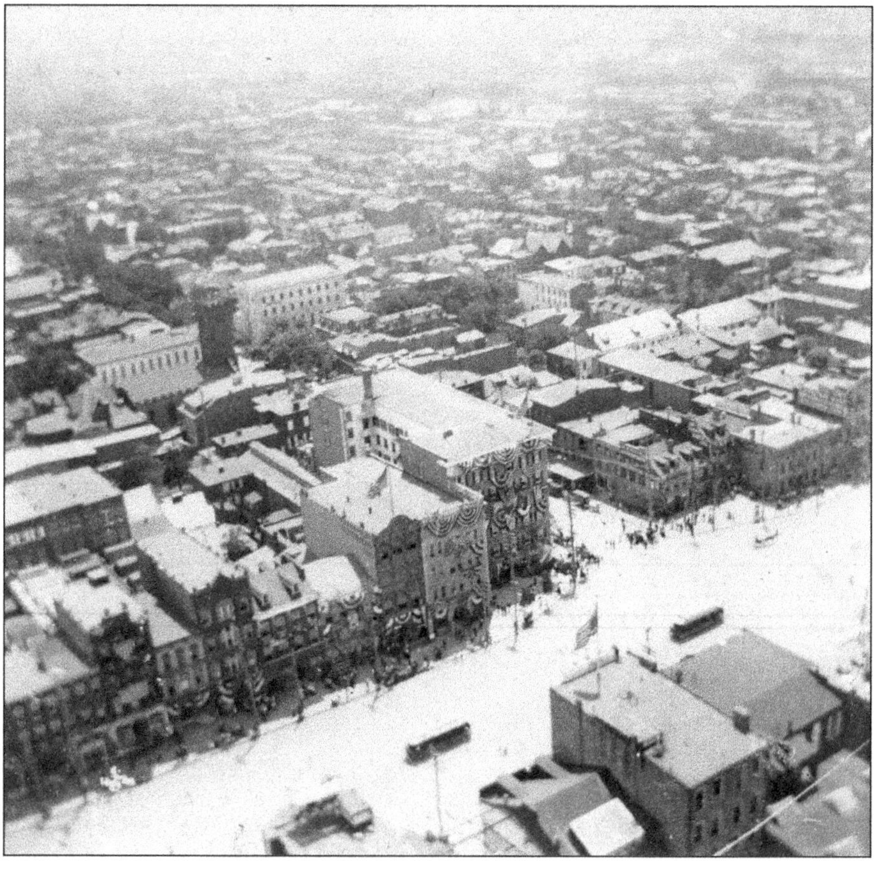

Two

Barnstorming and the Early Days

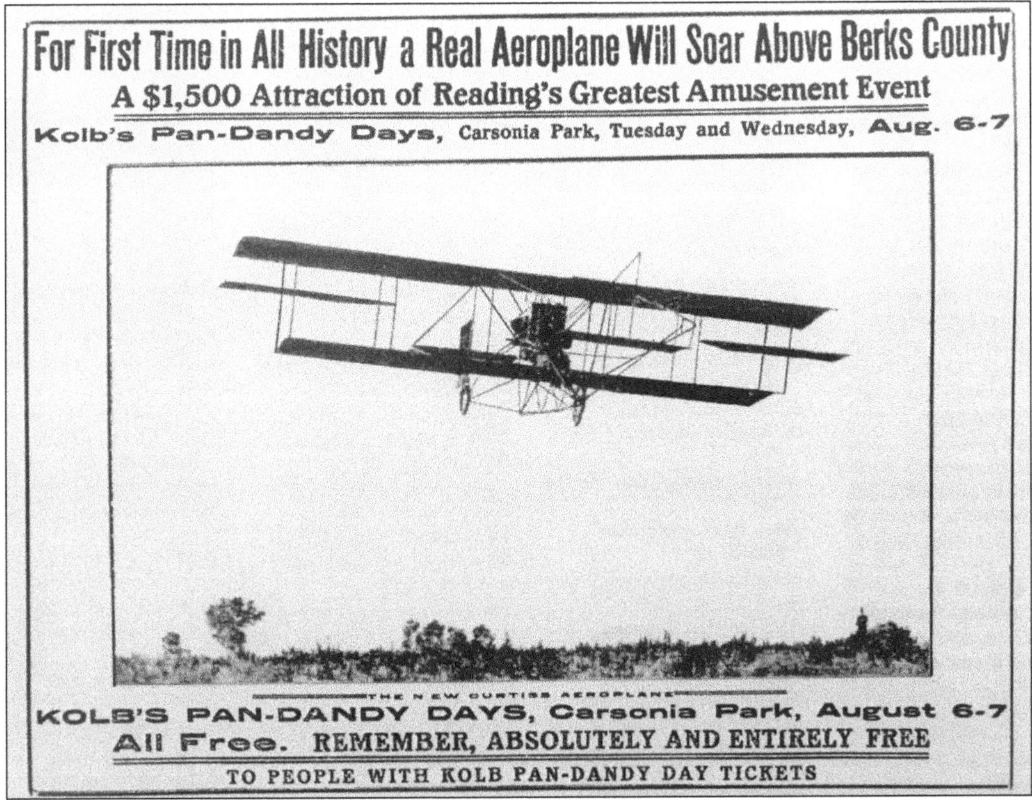

Kolb Bakery's Pan-Dandy Days was the first event to usher aviation to Berks County. In 1912, during the annual Kolb Bakery outing, a Curtiss biplane was the main event, performing flights from a baseball field. Frank C. Kolb, Reading's well-known businessman, expected a larger crowd than the previous year, but they exceeded the 1911 event by 10,000. On August 6 and 7, Charles F. Walsh thrilled the crowds with his flying. This image is from volume 23 of *The Passing Scene*. (Courtesy of George M. Meiser IX.)

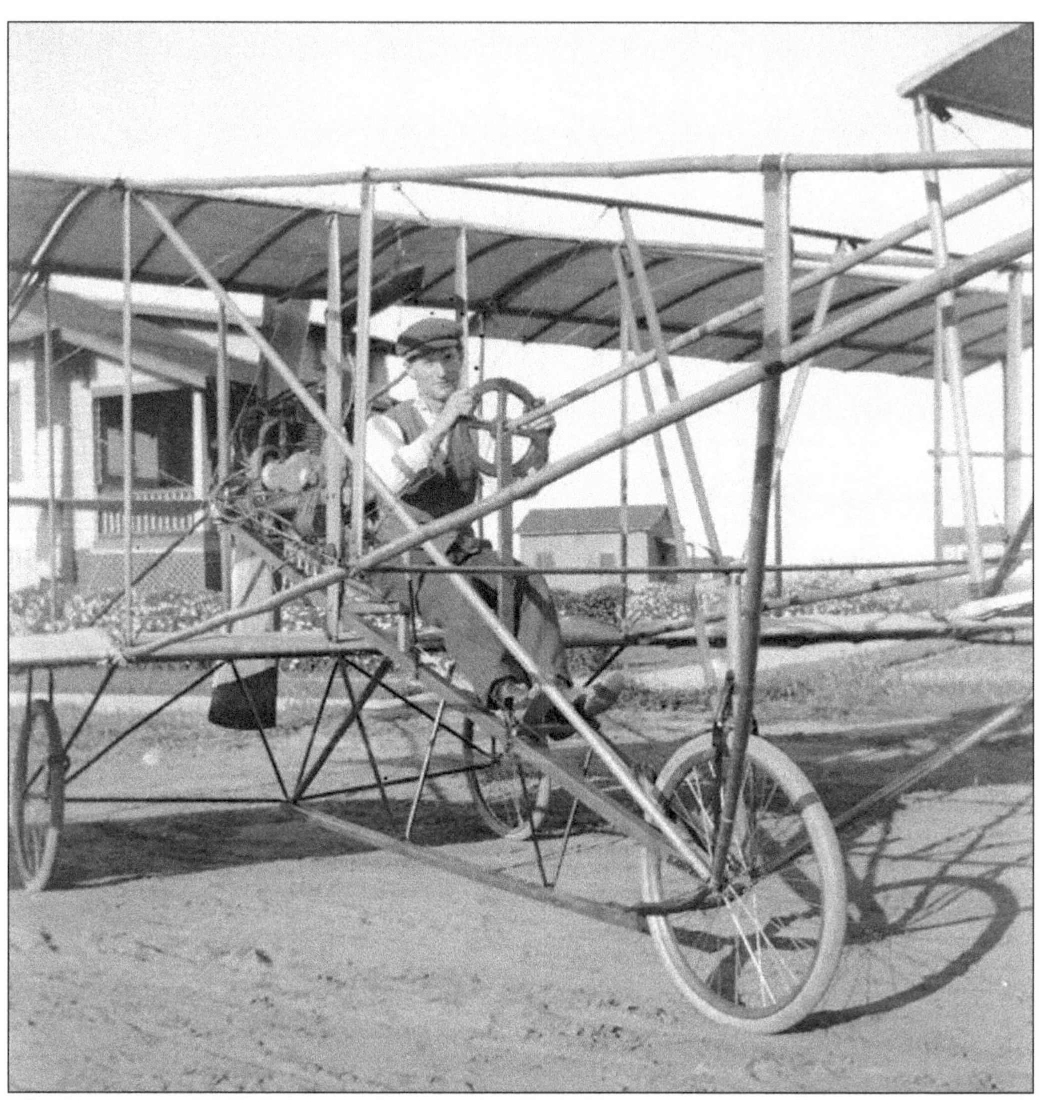

The first successful flight in Reading was conducted by San Diego aviator Charles F. Walsh on August 6, 1912, in a Curtiss biplane similar to the one shown here. The aviator conducted numerous flights over a two-day period on a baseball field opposite the Carsonia Park theater, soaring 2,500 feet above the ground for 18 minutes on the first flight of the day. Other flights included circling the city. As part of his grand finale, he rose 2,000 feet in the air and started his "spiral trick," which was once considered impossible to perform. Walsh was so popular during the promotional event that he was "carried about the grounds like a child," and everyone wanted to shake his hand. A squadron of state police had to be present at the event to preserve order. The event was so popular that extra trolley cars were required to transport the crowds. (Courtesy of Brian Deford.)

This statement, prepared by the Curtiss Exhibition Company, shows payment for flights at the 1912 Pan-Dandy Days and other national exhibitions in Springfield, Ohio, and Fon Du Lac, Wisconsin, during the first half of August 1912. Another contract stated that flights should only occur over the park to avoid potential injury to residents. Only two months later, on October 3, 1912, Charles Walsh died in front of 60,000 horrified people, including presidential candidate Woodrow Wilson, who watched him plummet to his death from an in-flight structural failure at 2,000 feet performing his spiral trick at the Trenton, New Jersey, fair. He was almost 38 years old. (Both, courtesy of National Air and Space Museum Archives.)

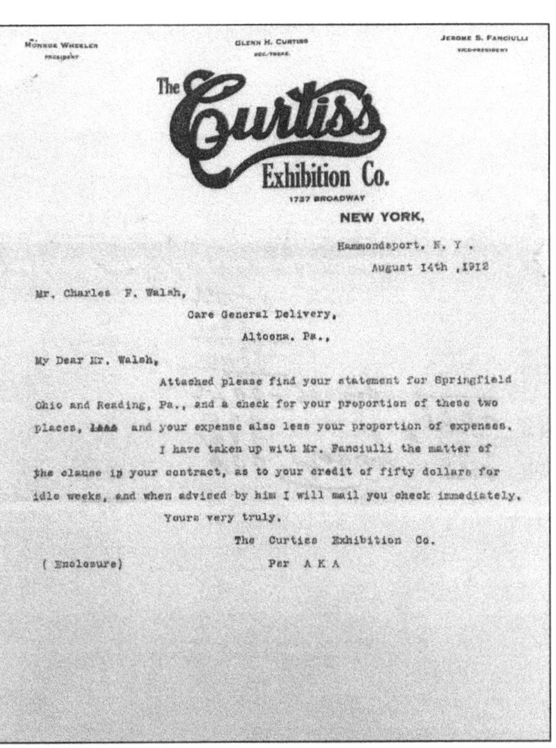

Monroe Wheeler, President — Glenn H. Curtiss, Sec.-Treas. — Jerome S. Fanciulli, Vice-President

The Curtiss Exhibition Co.
1737 Broadway
NEW YORK,

Hammondsport, N. Y.
August 14th, 1912

Mr. Charles F. Walsh,
Care General Delivery,
Altoona, Pa.,

My Dear Mr. Walsh,

Attached please find your statement for Springfield Ohio and Reading, Pa., and a check for your proportion of these two places, less and your expense also less your proportion of expenses.

I have taken up with Mr. Fanciulli the matter of the clause in your contract, as to your credit of fifty dollars for idle weeks, and when advised by him I will mail you check immediately.

Yours very truly,
The Curtiss Exhibition Co.
(Enclosure) Per A K A

C F WALSH STATEMENT from August 2nd to August 14th covering meets at Springfiled Ohio and Reading Pa.,

DEBITS.

August 2nd, 1912 Retained at Springfield Ohio	$ 300.00
August 2nd 1912 Parts shipped Fon Du Lac Wis.	27.75
August 7th 1912, 35% Expense Springfield Ohio	50.49
August 7th, 1912 35% Worthington Salary as Publicity man at Hamilton and Dubugue one half	7.88
August 7th, 1912 Account parts tools taken from Taylors bag at Hamilton (Note Credit)	3.09
August 7th 1912. Acct parts missing from machine when received from Hamilton and sent to Ellyson(Note Credit)	2.49
August 8th, 1912 Rental Springfield Ohio	780.00
August 9th 1912, Rental Reading Pa.,	585.00
August 9th, 1912, 35% expense Reading Pa.,	36.72
	$1,793.42

CREDITS.

August 7th 1912, Expense at Springfield , Ohio.	144.25
August 8th 1912, Proportion Springfield Ohio	1.200.00
August 9th, 1912 Proportion Reading Pa.,	900.00
August 9th, 1912 Expense Reading Pa.,	104.90
August 14th, 1912 Accounts parts credit by memo received from Mr Walsh,	2.49
August 14th, 1912 Accounts credit passed see letter from Mr Walsh	1.99
	$2,353.63

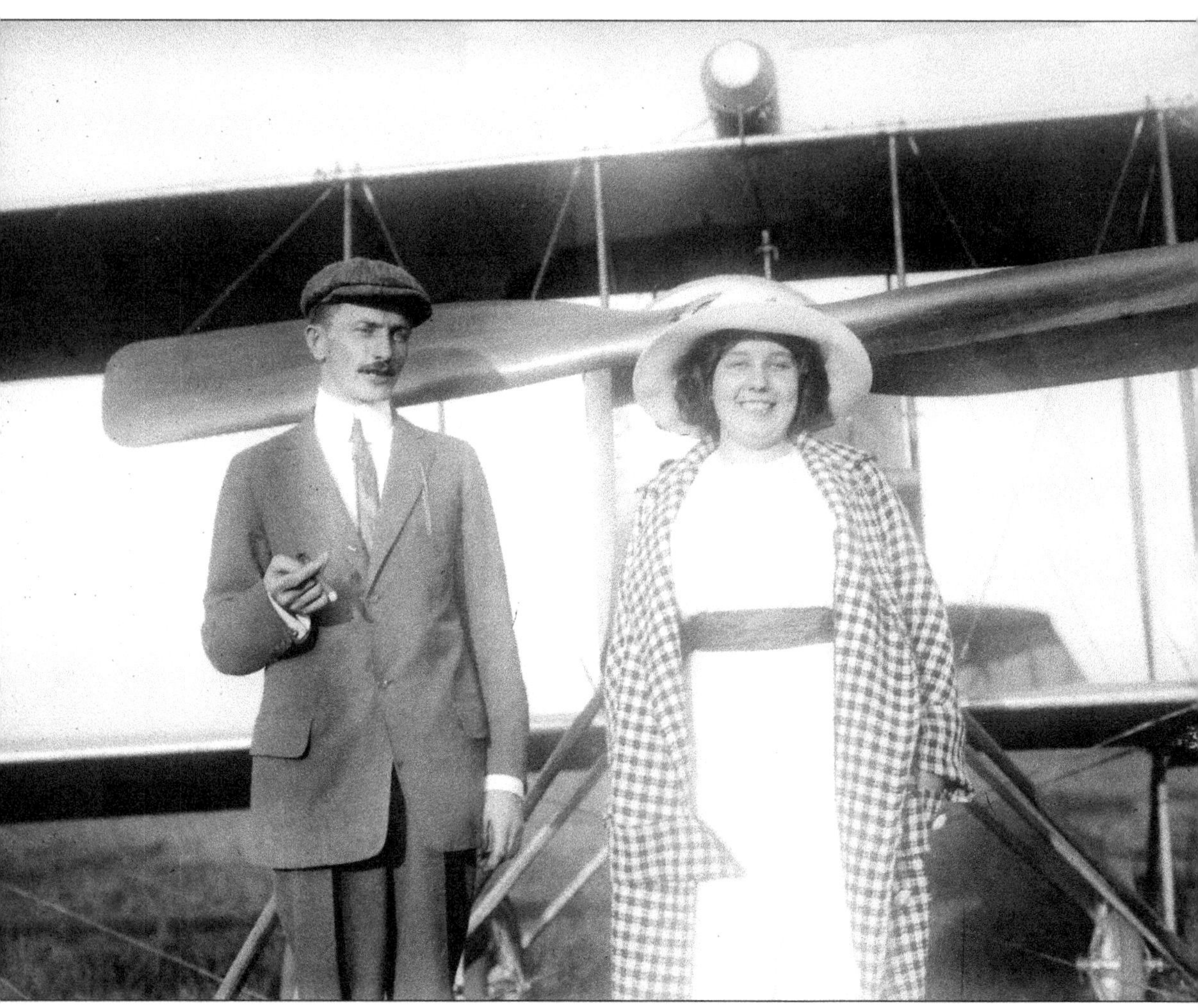

Following the first flight in Reading by Charles Walsh in August 1912, the Reading fairgrounds, located at the time in what is today City Park, became the scene of another flight on September 11, 1912. Joseph Richter of New York took off inside the fairground race track in his six-cylinder, 75-horsepower white Curtiss biplane in front of 1,000 people and circled the city before dropping a message box, attached to a parachute, at Fifth and Penn Streets for the Reading mayor. The exhibition was contracted between the Berks County Agricultural Society and a New York firm to demonstrate the practical use of aircraft. Also, an application was made with the local postmaster to deliver mail and souvenirs to Sinking Spring or Womelsdorf with all mail posted at the fairgrounds. Unfortunately, since the aviator did not arrive at the fair on September 10, as agreed upon with the postmaster, no air service was performed. (Author's collection.)

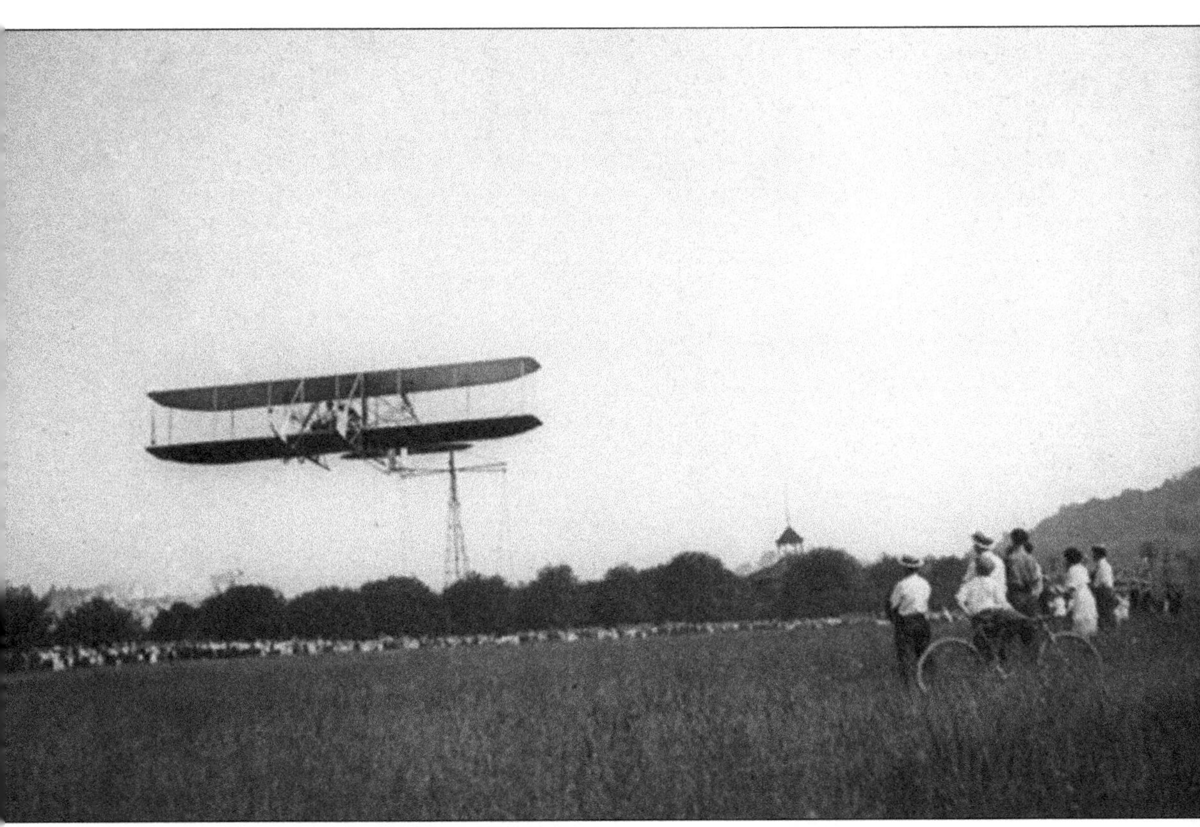

On July 9, 1914, a crowd of spectators gathered at Carsonia Park to witness two beautiful flights in a large, high-powered Wright biplane from the level lawn near the lake in Carsonia Park. The exhibition was staged by J.F. Berger Company, who held similar exhibitions in many of the large eastern cities. The aviator, a Mr. Hatts, did not attempt anything of an exceptionally daring nature. The only thrills were descents, swooping downward at a terrific speed in short spiral dips. Flying at approximately 50 miles per hour and rising steadily until a height of 500 feet was reached, the machine flew straight to Mt. Penn, cleared the summit, and then proceeded toward the city. The plane was visible from the city, and both flights were seen high above the mountains to the east. The Reading Transit Company ran cars on a three-minute schedule during the late afternoon and early evening to the park as an immense crowd was transported to the scene of the flights. (Courtesy of Joseph A. Webb.)

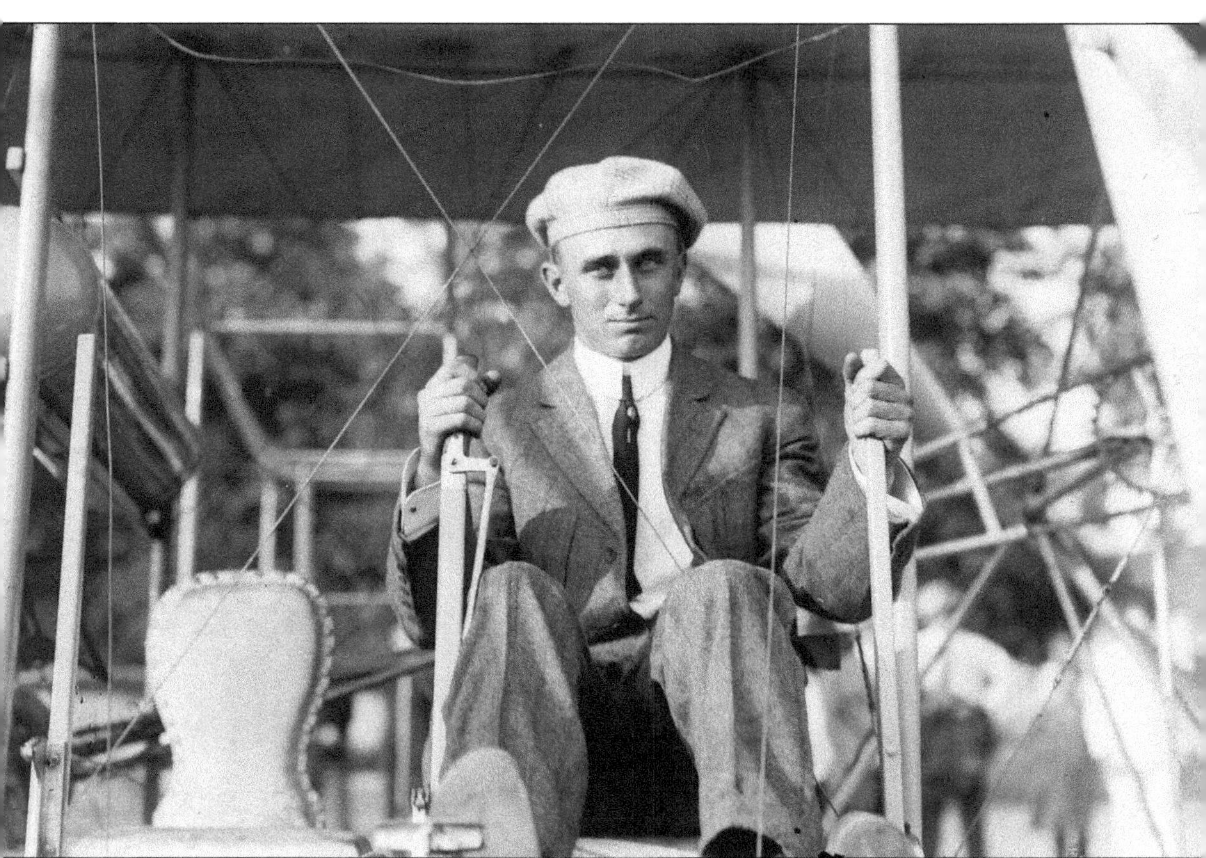

Harry Atwood was born in 1883 in Roxbury, Connecticut, and was trained by the Wright brothers before moving to Reading after he met Ruth L. Satterthwaite in Bar Harbor, Maine. Eventually, he married her on March 2, 1914, in the old Reading courthouse. During their courting, Ruth made news in Reading by becoming the first woman to fly in New Hampshire and was possibly the youngest woman to ever fly at that time. Ruth's parents were from Reading's high society. Her father, Alfred, was an investor and operator of the Reading Woolworth store on Penn Street. Harry, along with his wife, rented a house in Reading near the Sattherthwaite home. In 1914, Atwood considered flying across the Atlantic Ocean with his wife as part of a honeymoon trip but stated that regardless of her presence, nothing would interfere with his plan to make the flight. (Courtesy of Library of Congress.)

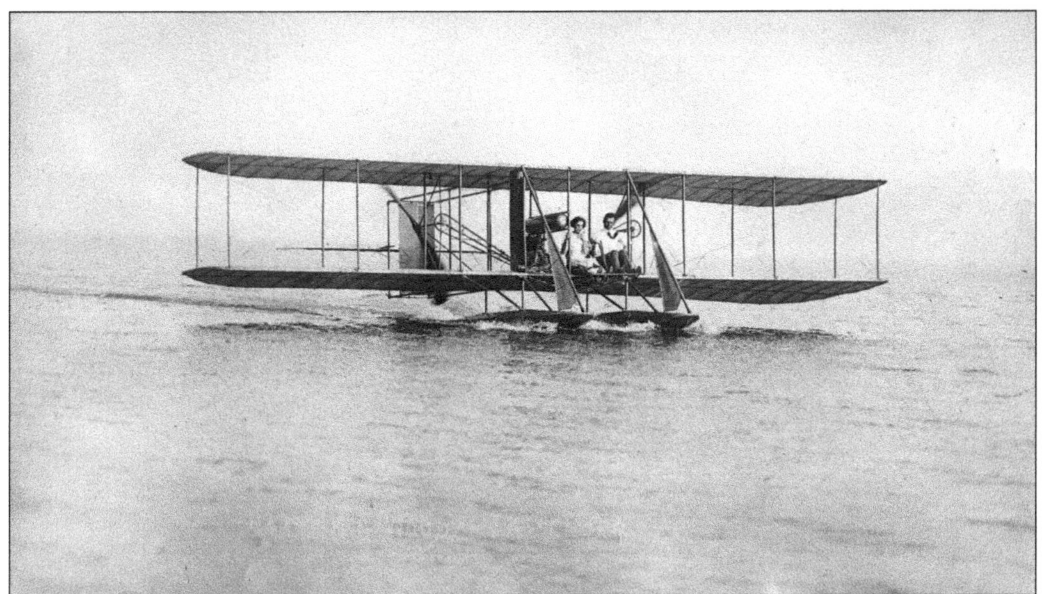

Harry Atwood built a large flying boat, shown here on Lake Erie in 1913, and eventually created the Atwood 12-180 V-shaped 12-cylinder water-cooled engine. His aircraft was bigger than the Curtiss flying boat *America*. Aircraft and engine development coincided with World War I. Since Reading's German population was significant, Reading's leaders chose not to financially back war production in Berks County. Eventually, the Atwood Aeronautic Company was opened in Williamsport in 1916. (Courtesy of Charles E. Frohman.)

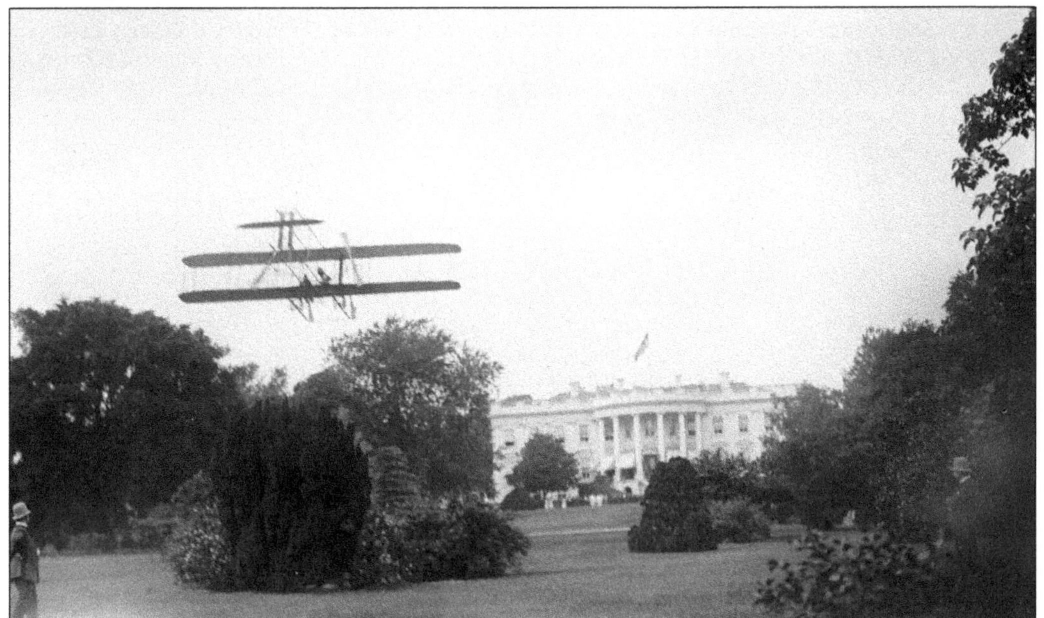

Harry Atwood was a flamboyant aviation pioneer and inventor known for landing his Wright Model B on the White House lawn on July 14, 1911, having lunch with President Taft and receiving a medal of accomplishment from the president. The biplane stopped within 25 feet of the president, and as Atwood sat, resting for a moment before leaving the airplane, spectators crowded around him. (Courtesy of Library of Congress.)

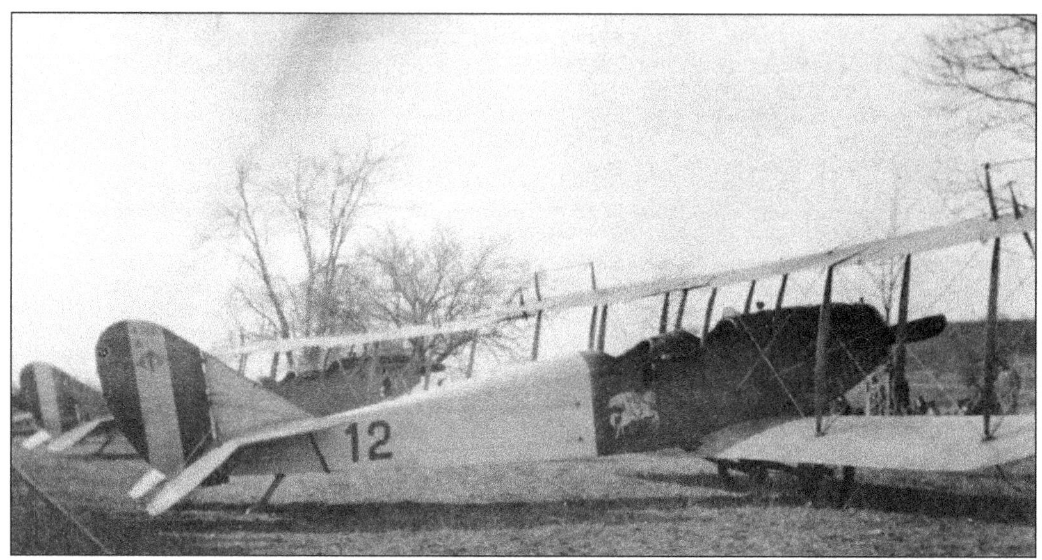

During the late 1910s to mid-1920s, numerous airplanes landed on fields behind Shillington High School adjacent to the Berks County Poorhouse buildings bordering Lancaster Avenue. Bigony Field was on Lancaster Avenue, but the Poor House field, composed of 100 acres of flat land, was more suitable for aircraft during the early days of barnstorming. Eventually, many aviators and city leaders believed the Poor House field was the ideal location for the city's first municipal airport because of its proximity to the city and its easy access by trolley. In 1927, requests were still being made to develop an airfield when Arthur Arrowsmith, a Kutztown resident and aircraft owner, challenged the county to sell 20–30 acres of the Poor House field for this use—a request that never came to fruition. (Both, courtesy of Barry Nelson.)

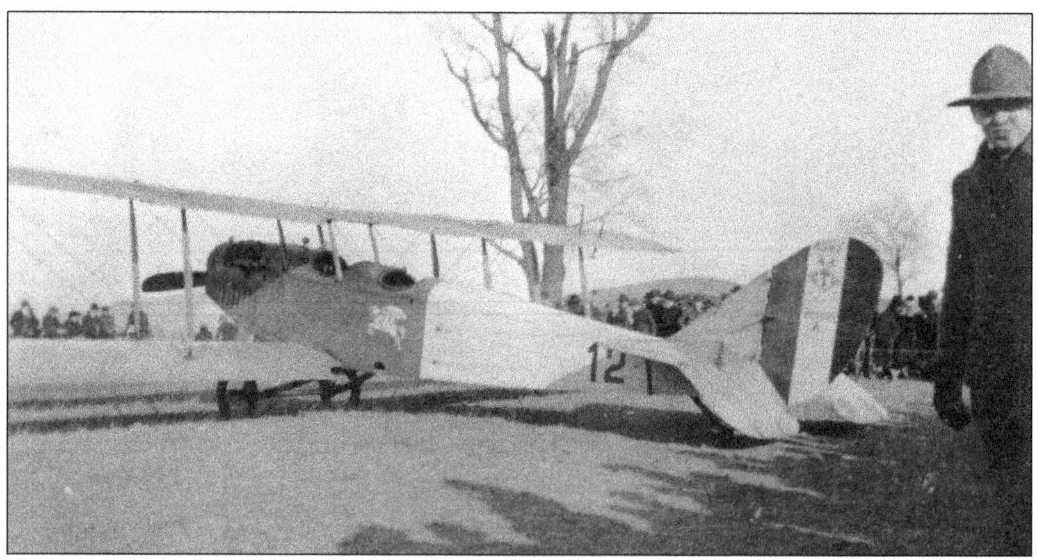

Three

Berks County Pioneer Aviators

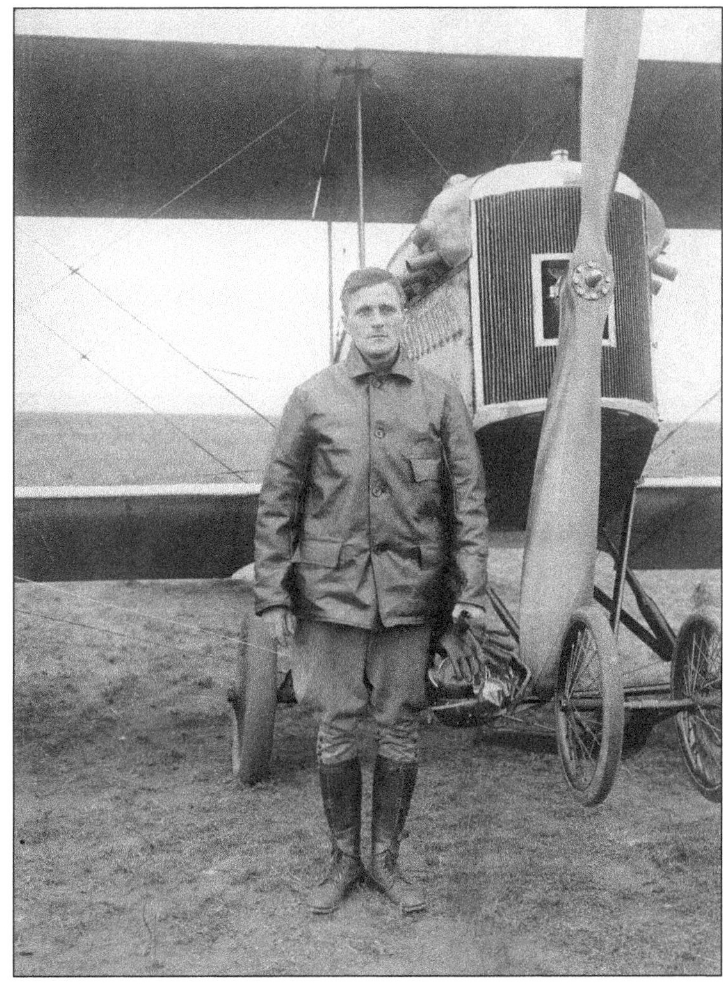

Shown here in 1916 is one of Berks County's best-known aviation pioneers, Carl Spaatz. Born June 28, 1891, in Boyertown, he graduated from West Point in 1914 and learned to fly in San Diego, receiving his wings in June 1916 and assigned to the First Aero Squadron in Columbus, New Mexico. The Reading Municipal Airport was dedicated Gen. Carl A. Spaatz Field on August 12, 1951. (Courtesy of Sime Bertolet.)

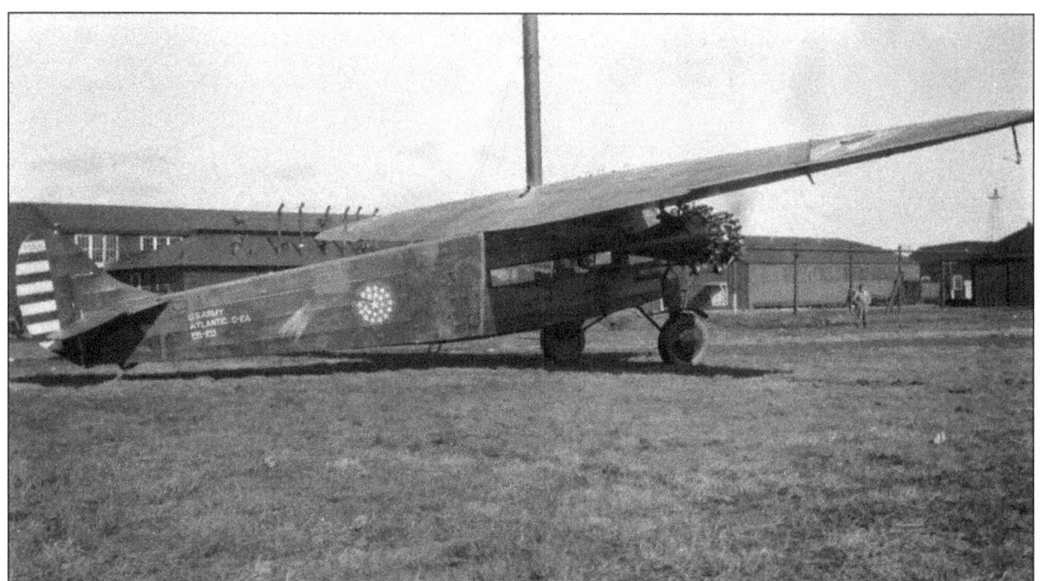

On January 7, 1929, a Fokker C-2 named *Question Mark* and a crew under the command of Maj. Carl Spaatz (third from right below), set a flight endurance record of 150 hours and 40 minutes using air-to-air refueling. Although the process was previously attempted, Spaatz and his crew perfected the technique over 43 times, with nine refuelings made at night, and covering a distance of 11,000 miles before landing. Unfortunately, the Army was not impressed by the accomplishment and did not provide further funding. Years later, a Curtiss Robin aircraft broke the record, staying airborne for 653 hours and 34 minutes. After World War II, once the military realized the advantages of in-air refueling, Spaatz made it a priority for the military, and in 1953, the process was implemented in Air Force squadrons. (Both, courtesy of Terry Sroka.)

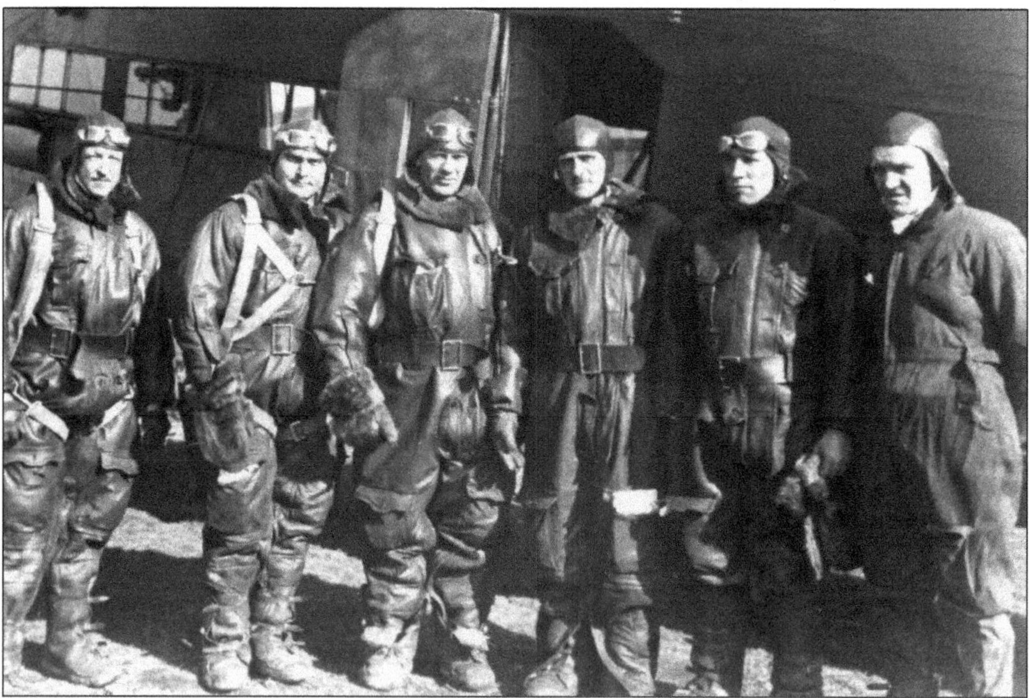

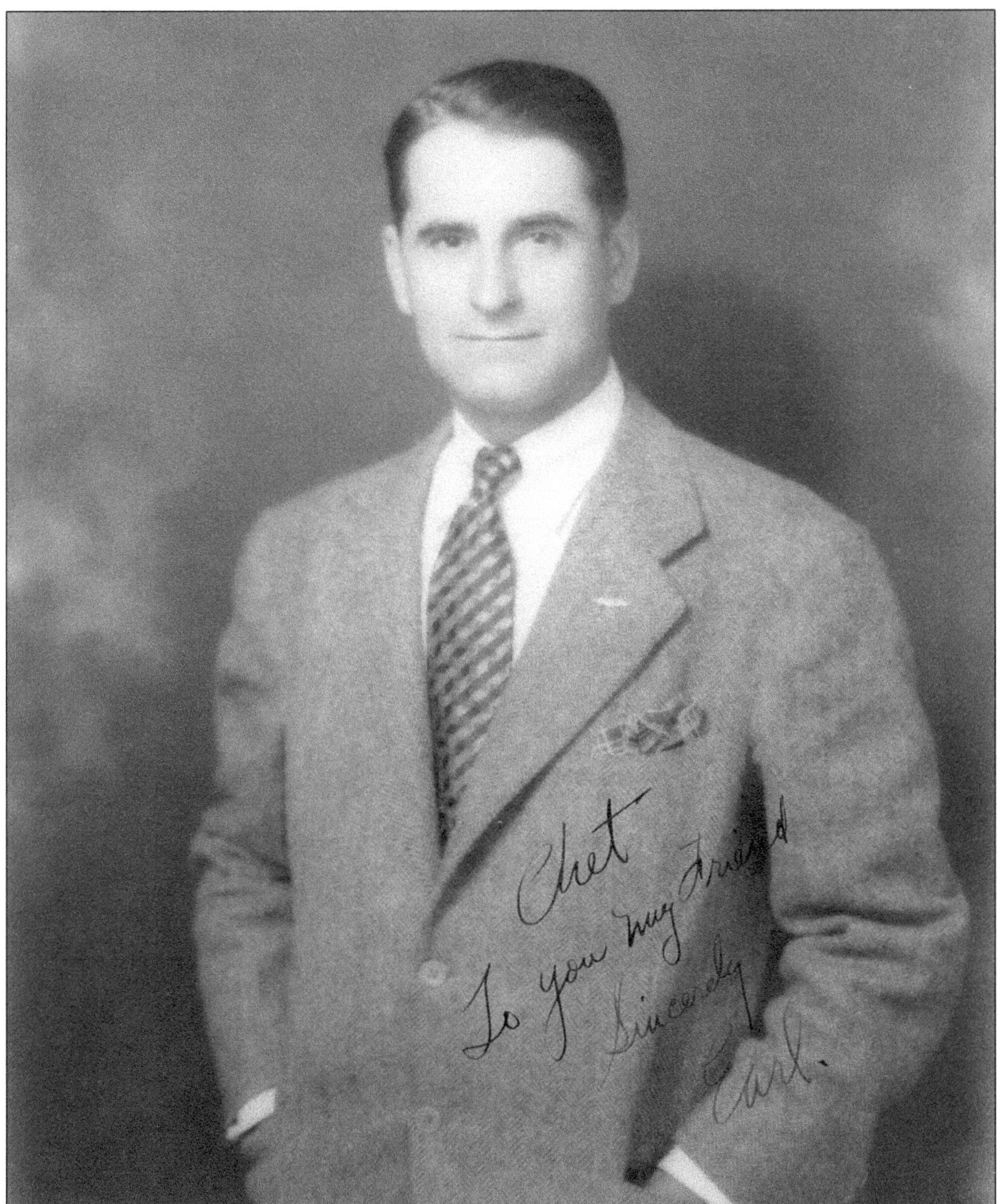

J. Earl Steinhauer was an important figure in American aviation history. Born in Shillington, Pennsylvania, in 1897, Steinhauer became a pilot in 1926 and held pilot license No. 237, signed by Orville Wright. Steinhauer was in charge of security and the refueling of the *Spirit of St. Louis* while working at Roosevelt Field during Charles Lindbergh's takeoff on his historic flight to Paris in 1927. Another contribution Steinhauer made to aviation was proposing a "National Aviation Day," which was established in 1939 by Pres. Franklin Roosevelt. He possessed an inventive mind, creating an aviation product still used by pilots today—the control lock. The control lock ensures that winds do not damage an airplane's control surfaces while exposed to external elements. Simple in design, the control lock attaches to the aircraft's aileron or rudder to ensure it cannot be moved by wind. (Courtesy of Don Warrington.)

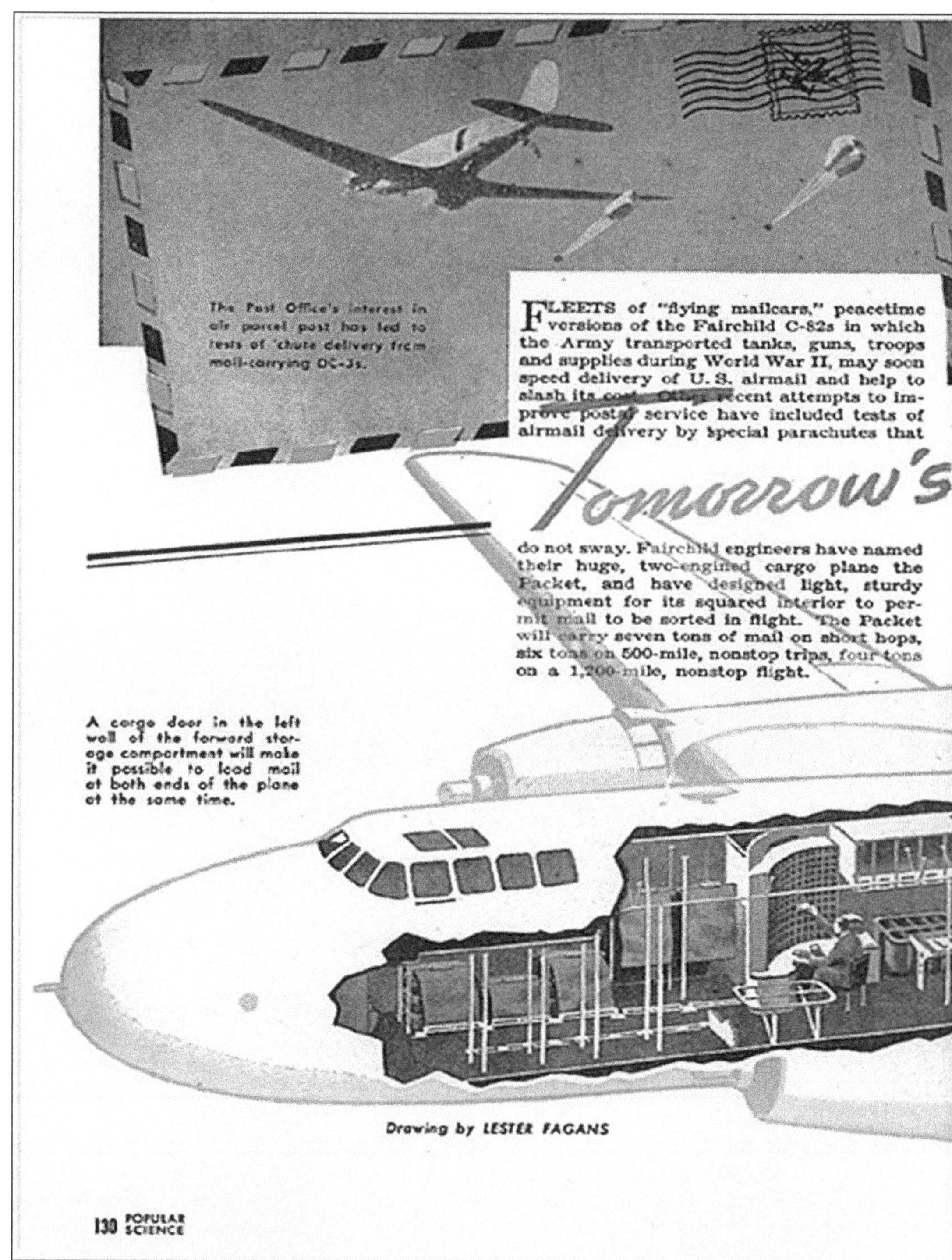
FLEETS of "flying mailcars," peacetime versions of the Fairchild C-82s in which the Army transported tanks, guns, troops and supplies during World War II, may soon speed delivery of U. S. airmail and help to slash its cost. Other recent attempts to improve postal service have included tests of airmail delivery by special parachutes that

Tomorrow's

do not sway. Fairchild engineers have named their huge, two-engined cargo plane the Packet, and have designed light, sturdy equipment for its squared interior to permit mail to be sorted in flight. The Packet will carry seven tons of mail on short hops, six tons on 500-mile, nonstop trips, four tons on a 1,200-mile, nonstop flight.

The Post Office's interest in air parcel post has led to tests of 'chute delivery from mail-carrying DC-3s.

A cargo door in the left wall of the forward storage compartment will make it possible to load mail at both ends of the plane at the same time.

Drawing by LESTER FAGANS

130 POPULAR SCIENCE

J. Earl Steinhauer also held other significant positions within the aviation industry such as operations manager of Washington National Airport in 1941. His position at Fairchild Aircraft in Hagerstown, Maryland, led to the October 1, 1947, development of the "Flying Mail Car," as featured in this June 1946 edition of *Popular Science* magazine, in which a C-82 Packet was converted

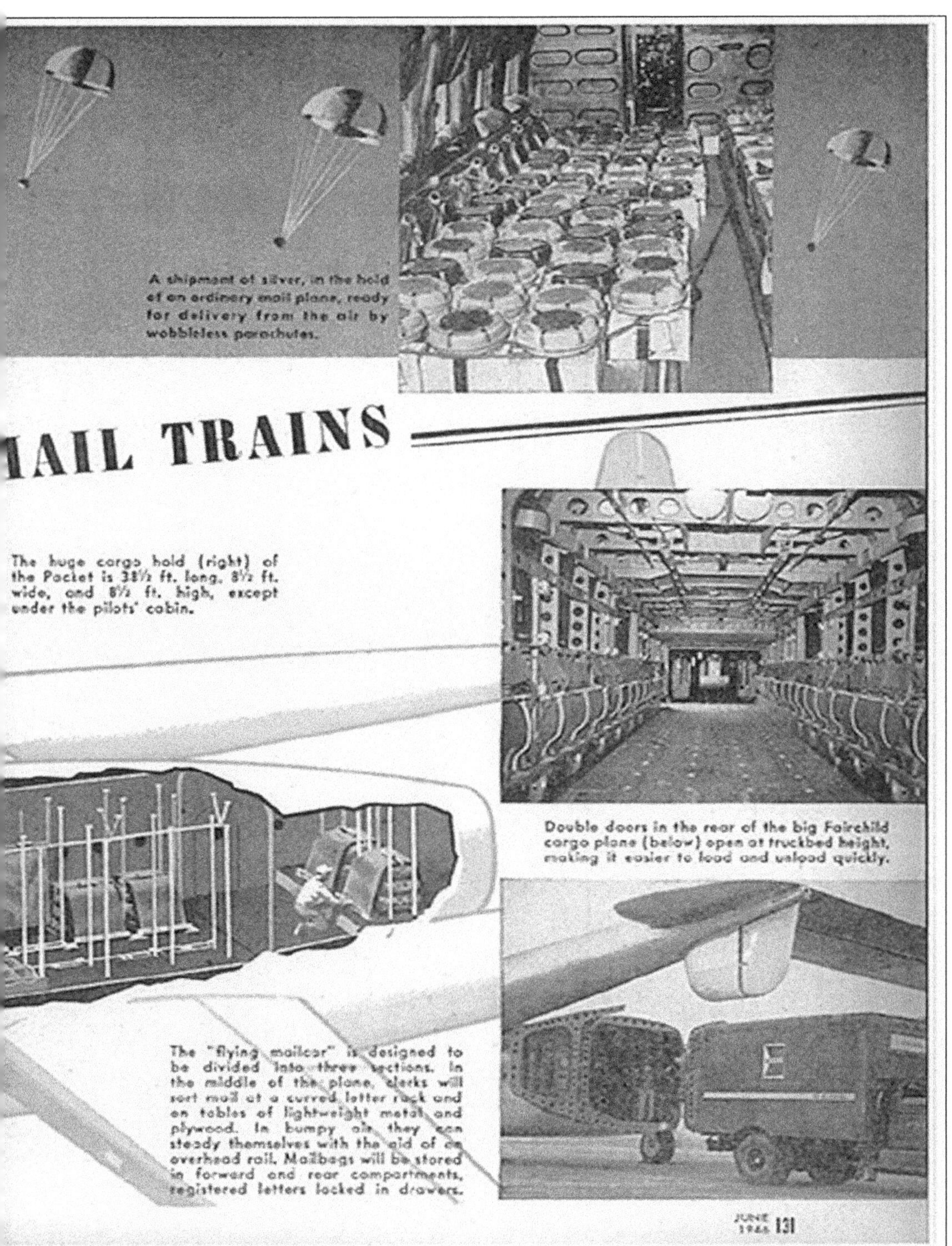

A shipment of silver, in the hold of an ordinary mail plane, ready for delivery from the air by wobbleless parachutes.

IAIL TRAINS

The huge cargo hold (right) of the Packet is 38½ ft. long, 8½ ft. wide, and 8½ ft. high, except under the pilots' cabin.

Double doors in the rear of the big Fairchild cargo plane (below) open at truckbed height, making it easier to load and unload quickly.

The "flying mailcar" is designed to be divided into three sections. In the middle of the plane, clerks will sort mail at a curved letter rack and on tables of lightweight metal and plywood. In bumpy air, they can steady themselves with the aid of an overhead rail. Mailbags will be stored in forward and rear compartments, registered letters locked in drawers.

JUNE 1946 131

into a post office on wings. This marked the beginning of 5¢ airmail service in the United States with its initial flight from New York's La Guardia airport to San Francisco, California. The name "Packet" originated from the packet boats that hauled mail, passengers, and freight in Europe. (Courtesy of Purdue University Archives.)

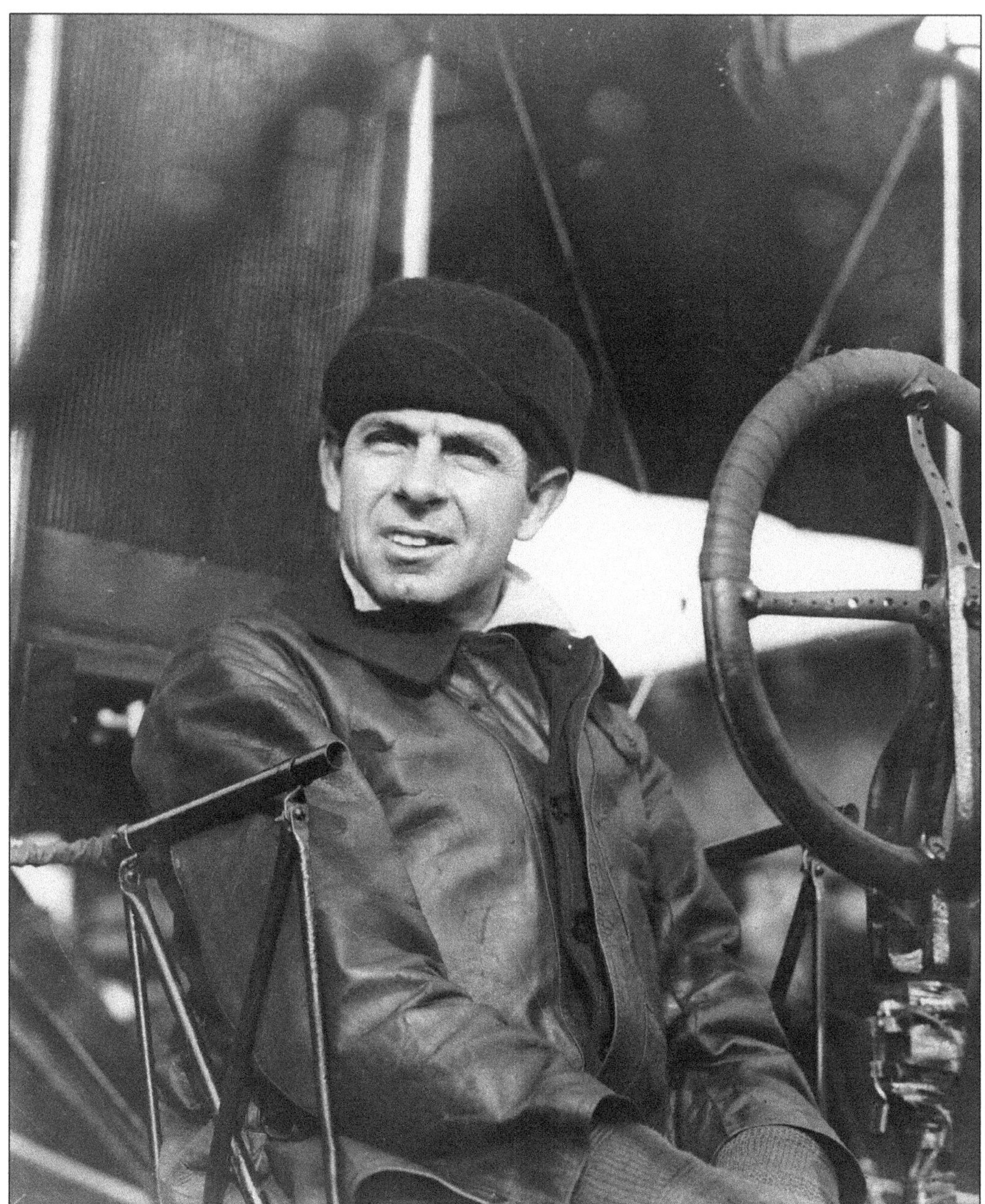

Clifton O. Hadley, pictured at Whander Field, was born in Ohio on February 10, 1876, then resided in Tarrytown, New York, for a number of years before moving to 1034 Washington Street in Reading in 1924. He was one of the first men to fly an airplane and flew one of the first airplanes made in America on and over the frozen Hudson River on February 12, 1912. Other firsts claimed by Hadley were being the first man ever sworn in as an airmail pilot in 1911 and becoming the first man to be shot down in an airplane—by a stray bullet from a hunter's gun, while Hadley was attempting to beat that year's flight endurance record. In 1935, he flew his Monocoupe around Reading and landed on the ice of the Hudson River again to celebrate his 58th birthday before returning to Reading. (Courtesy of National Air and Space Museum.)

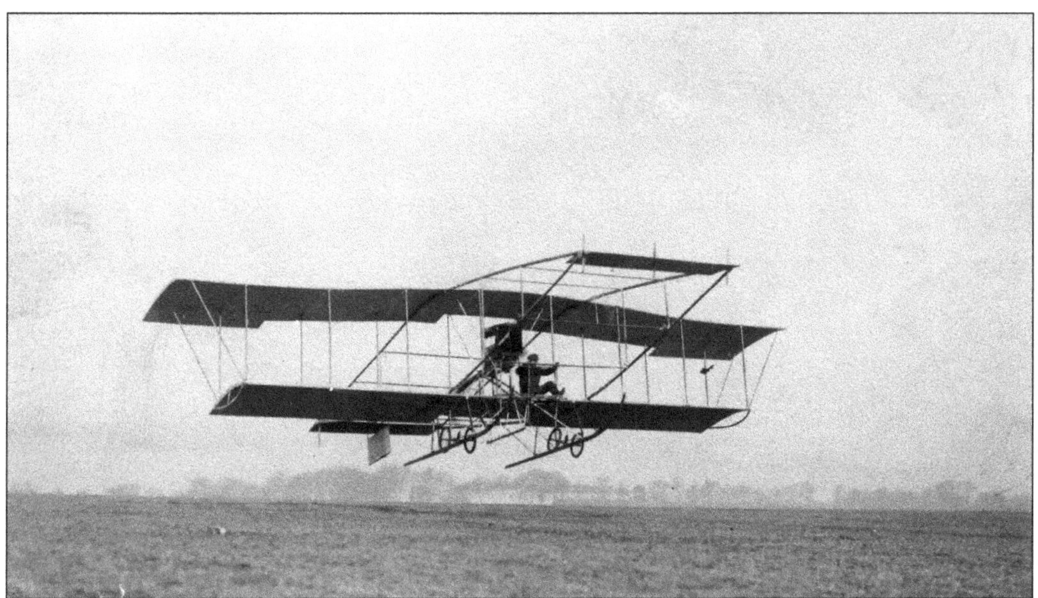

Clifton O. Hadley began his aviation career when he taught himself to fly at the Empire City Race Track property in Mineola, New York, on October 6, 1910. The aircraft, having a 45-foot wingspan, was designed and built by Hadley and a partner in Virginia in 1909. The "Hadley and Blood" machine had a unique feature that Hadley invented for varying pitch control. (Courtesy of National Air and Space Museum.)

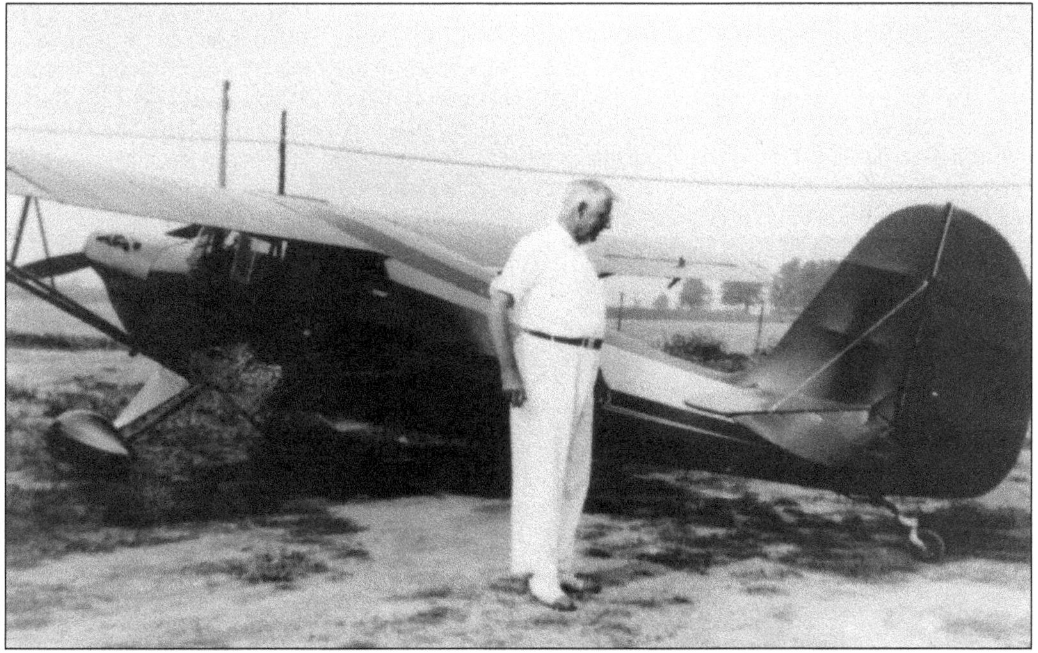

Hadley, pictured at Whander Field, established an interior decorating business in Reading. He was aeronautical advisor to the Reading Airport Commission, contributing to the design of the Reading Municipal Airport. Additionally, he led the Reading Aero Club by acting as secretary-treasurer, was technical advisor at the new Reading Municipal Airport, and bicycled 2,700 miles from Portland, Oregon, to Columbus, Ohio, in 1896. (Courtesy of National Air and Space Museum.)

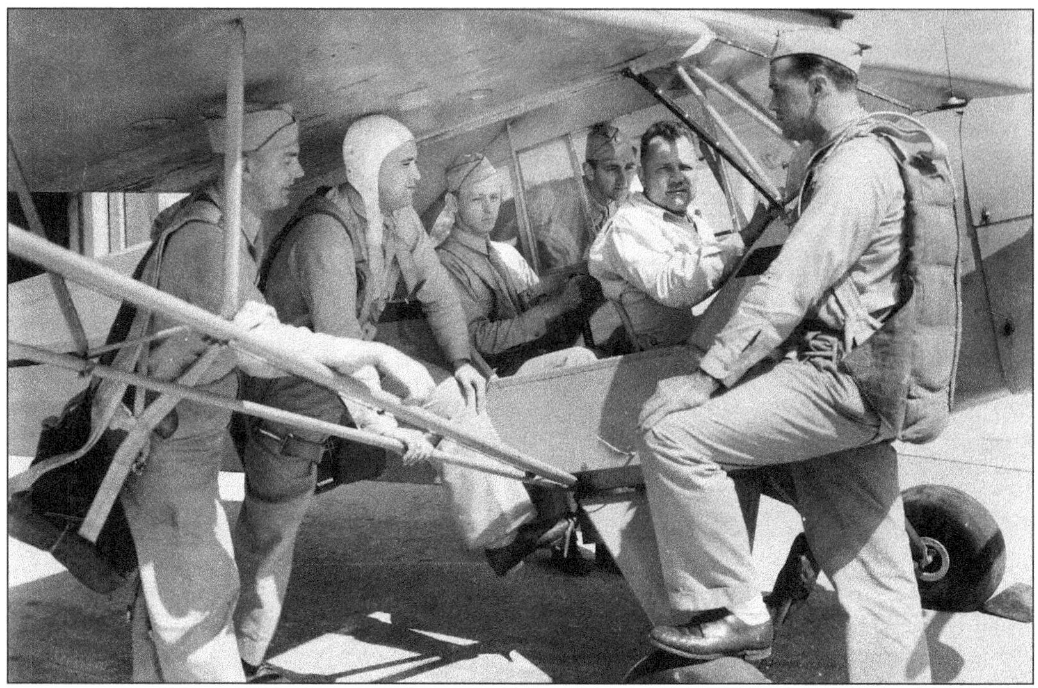

Marquis C.J. Markle (above, seated in front seat among Albright College cadets, and below, demonstrating hand propping techniques at the Reading Army Airfield in 1943) was known as the first pilot to own an airplane in Berks County. He operated an airfield in Pennside and established the first regular flight services between Reading and Philadelphia on September 8, 1927. He was also an integral part of Atlantic City's Steel Pier entertainment program by being the pilot for a daredevil parachutist called the "Human Rocket." During World War II, Markle actively trained more than 2,000 pilots for the Army, Navy, Marine Corps, and the Albright College Civilian Pilot Training Program as a flight instructor for the former Reading Aviation Service. Other contributions he made to the war effort were flying bombing study missions for the development of the Norden bombsight. (Both, courtesy of Albright College.)

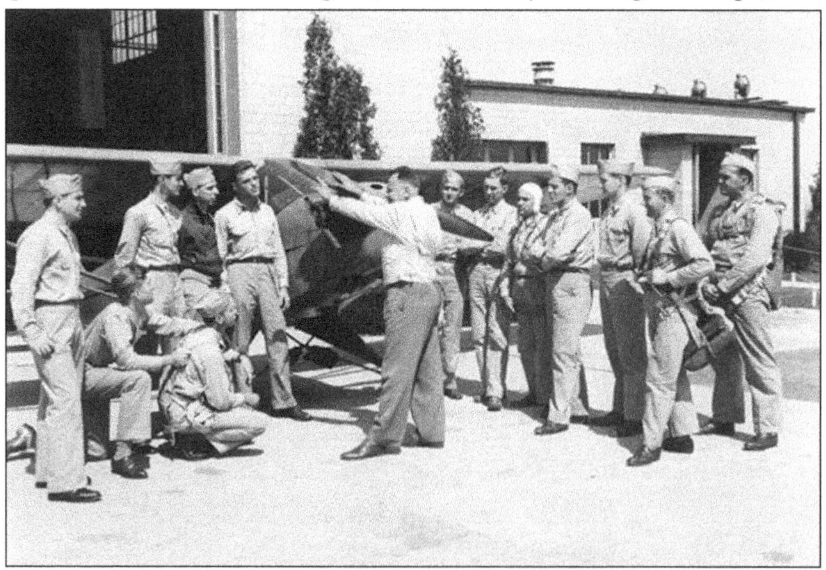

Eddie Nibur (left), in front of a New Standard with Quinton Cudney at Madeira Field in 1935, was an extremely accomplished pilot and a significant contributor to Berks County's aviation history. After high school, he took a job in a broker's office, but felt it was too tame for him; so, he studied aeronautical engineering, received flight instruction by nationally known inventor, engineer, and aviator Assen Jordanoff, and earned his commercial license on September 8, 1928. (Courtesy of Paul Nibur.)

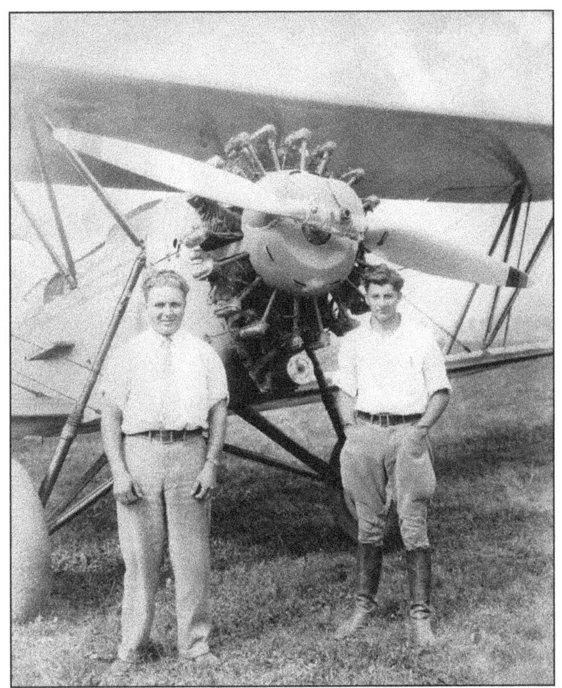

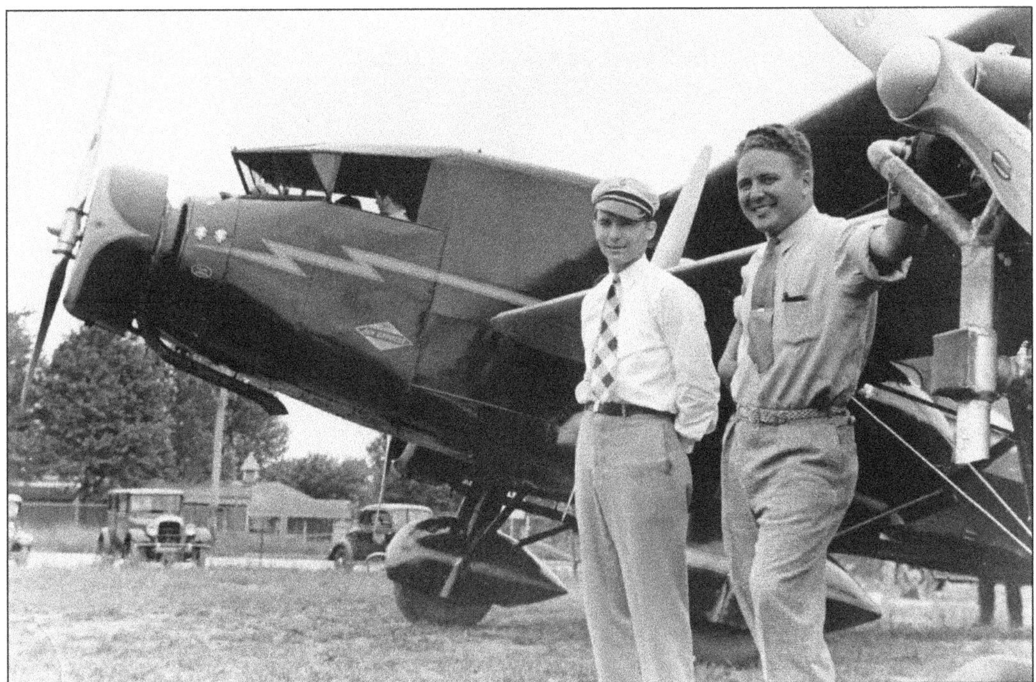

In 1934, Eddie Nibur moved from Teterboro to Reading after signing an agreement with farmer Charles Madeira. He brought his team that included Quinton Cudney (pictured here at left), Joe Levan (a Reading friend), Hugh Sullivan (promotion), and Charlie Douglass (helper). Nibur expanded Madeira Field by grading and lengthening the runways that extended east to west from Kutztown Road to Fifth Street Highway and south to north beginning at George Street. (Courtesy of Paul Nibur.)

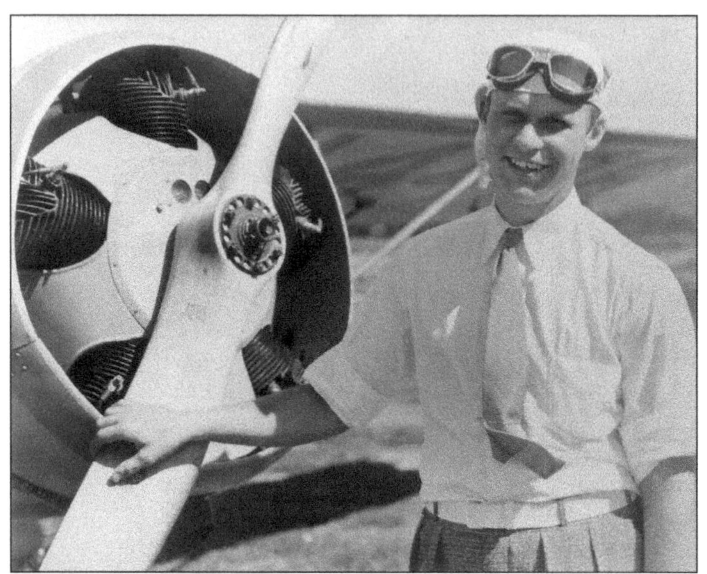

The Wyomissing Industries publication the *Yarn Carrier* featured this picture of Alfred M. "Sime" Bertolet during the 1930s. Eventually, he became a local aviation pioneer and founder of Reading Aviation Service, Suburban Airlines, Public Aviation, and the Reading Airshow. He discovered a love for flying as a young boy walking past Whander Field in Leesport. In 1970, he received the governor's aviation trophy for his contributions to aviation in Pennsylvania. (Courtesy of Sime Bertolet.)

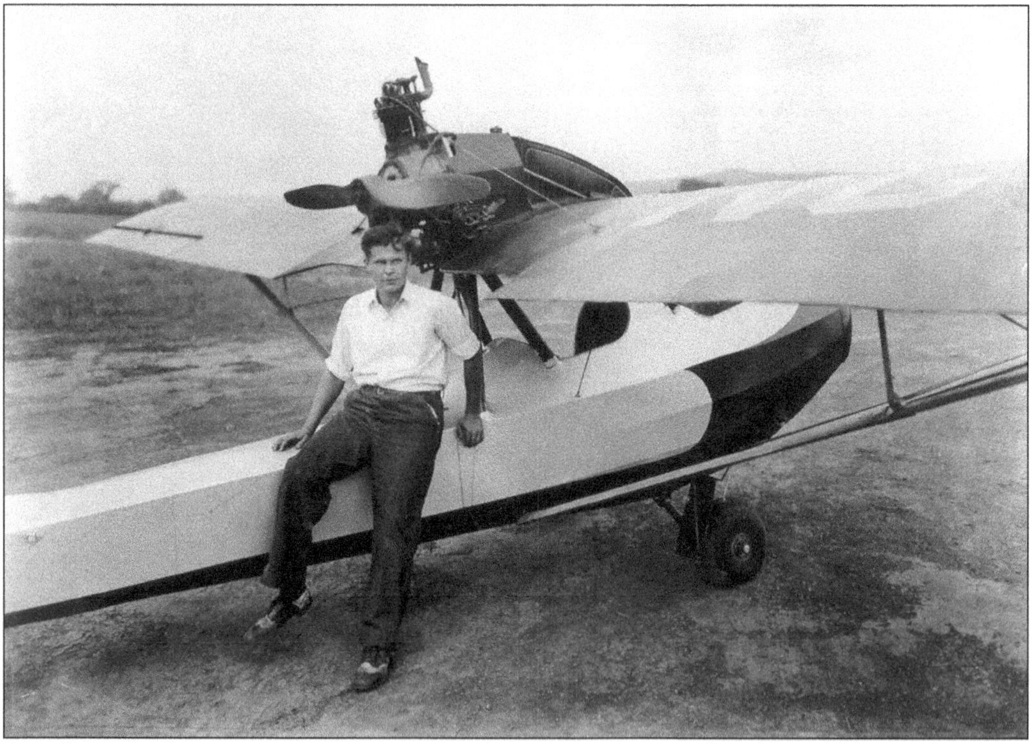

Bertolet is seen leaning against a Curtiss-Wright CW-1 Junior at Whander Field during the 1930s. The CW-1 Junior, originally named the CR-1 Skeeter, was a light sports aircraft produced during the 1930s that had a capacity of one passenger and a range of 200 miles. Sime soloed by the time he was 18 after receiving free flight lessons for work performed at Whander Field. (Courtesy of Sime Bertolet.)

One of the first licensed flight instructors in the area and a significant contributor to commercial aviation in Berks County was R. Harding "Breity" Breithaupt, shown here at a 1960s Reading Airshow. Breity (left) raised capital with partner Sime Bertolet (right) and bought a company, naming the new enterprise Reading Aviation Service. The business was interrupted when both men joined the military, but resumed after World War II. (Courtesy of Sime Bertolet.)

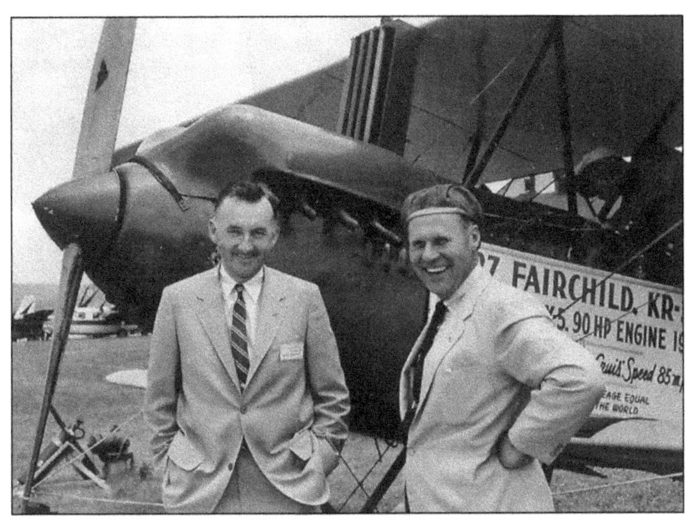

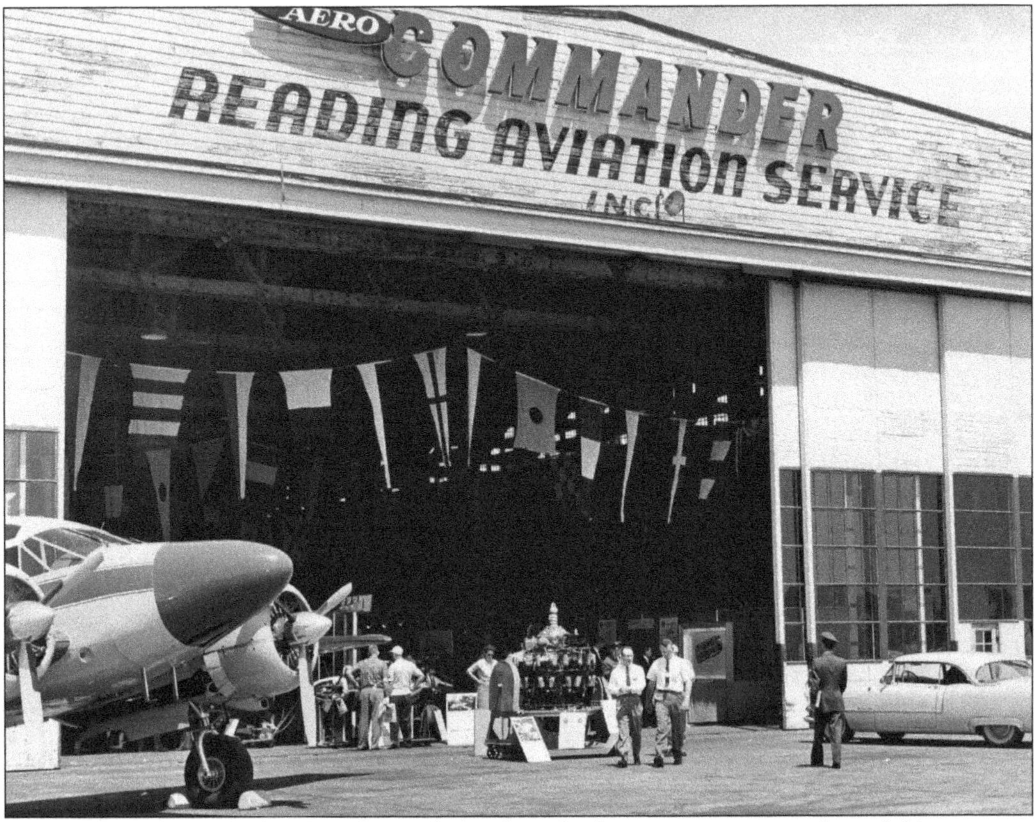

As vice president of sales, Breithaupt helped guide the company into a formidable aviation operation known nationwide for its stellar services. Reading Aviation Service spawned Reading Airlines, becoming Suburban Airlines in 1978. The company's annual maintenance and operations meeting became known as the Reading Airshow and grew in size and scope to eventually rival the biannual Paris Air Show. Pictured is a ramp display at the June 1961 show. (Courtesy of the National Air and Space Museum.)

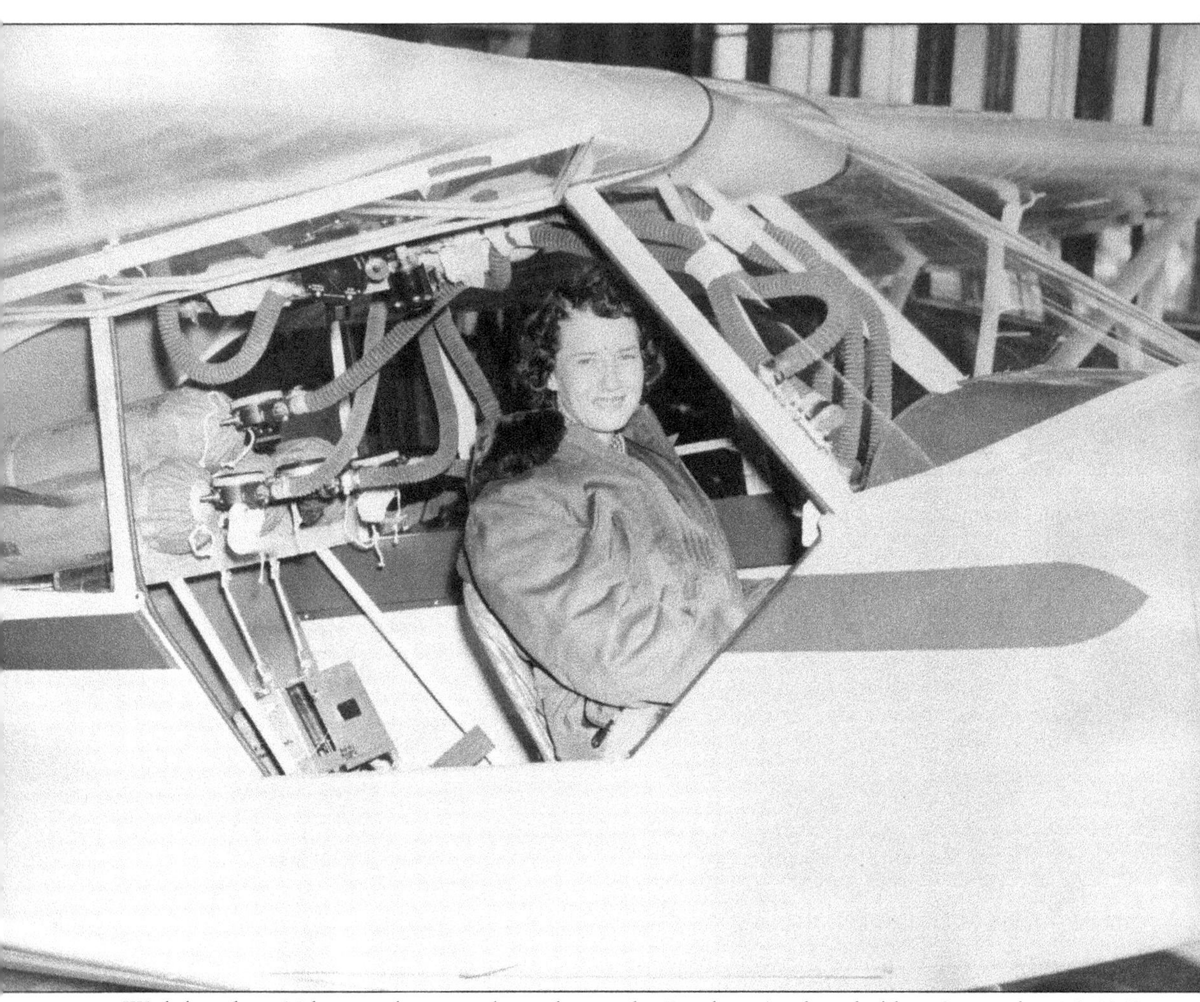

With less than 30 hours piloting a plane, during the Reading Airshow held on September 26, 1949, Mildred Zimmerman flew her Piper Cub Special to a world-record altitude of 26,138 feet—nearly five miles above the earth's surface—in front of a crowd of 40,000 people. After an estimated flight time of four hours and five minutes, she landed and rolled to a stop in front of a cheering crowd. The sealed barograph was taken to Washington for examination and calibration. Nine years prior to this event, another Berks County resident, 20-year-old Betsy Ross of Stouchsburg, attempted the same feat, reaching an unofficial altitude of 18,200 feet only to find that the barograph installed by the National Aeronautics Association was defective and registered an altitude of only 5,000 feet. Shortly afterward, 20-year-old Grace Huntington of Pasadena achieved and held the record of 24,310 until Zimmerman smashed it. To help her prepare for the flight, the Army and National Guard fitted the plane with oxygen bottles and hoses and loaned Zimmerman an Arctic flying suit to handle the harsh high-altitude environment. (Courtesy of Cathryn Appel.)

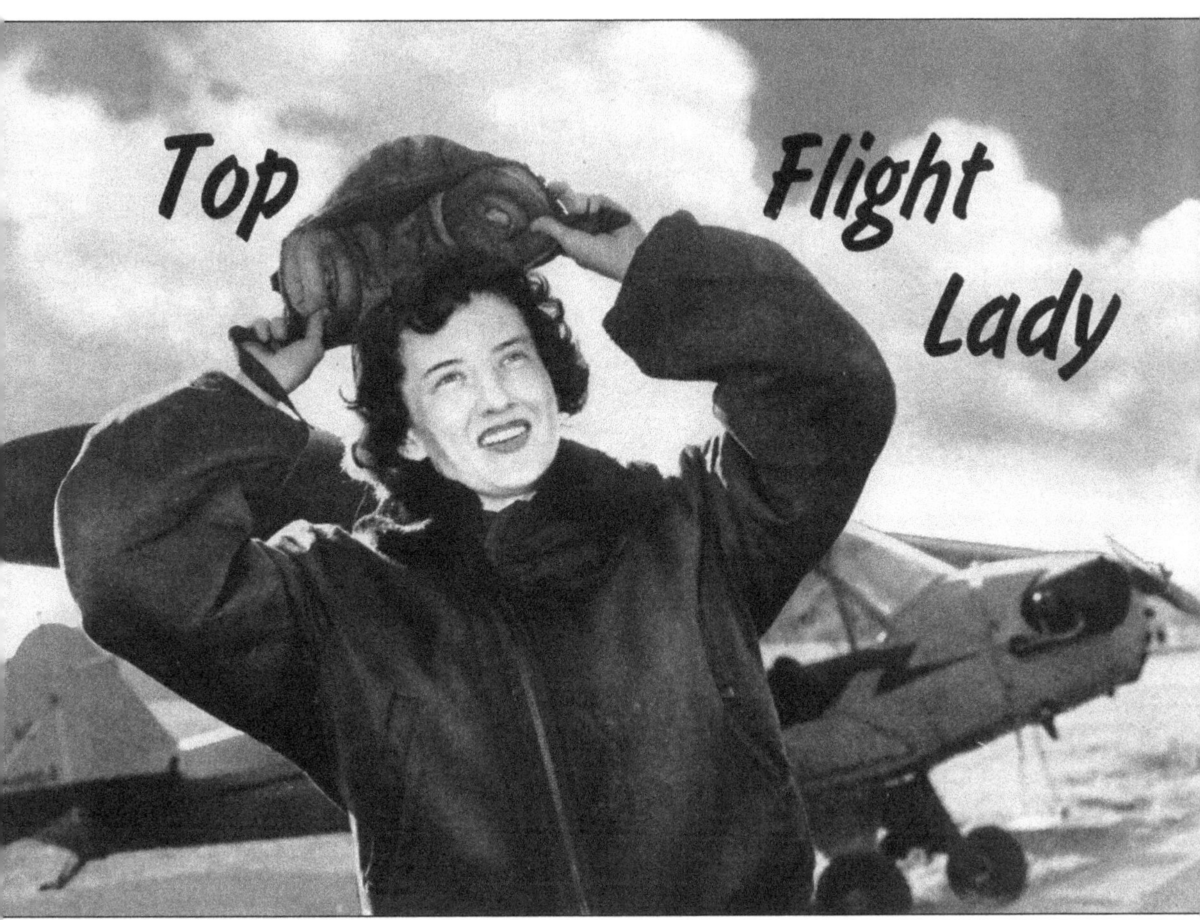

Shown in front of her non-supercharged Piper Cub Special, Mildred Zimmerman was a feature story in a 1950 Cities Service publication. Cities Service was later renamed Citgo, an oil and gas company. The same picture was featured in the October 23, 1949, edition of the *Philadelphia Inquirer Magazine*. Zimmerman, of Reading, began taking flying lessons at Schuylkill County Airport when she fell in love with another aviator, Reed Zimmerman, while in high school. As a graduation gift, she received an airplane and eventually married Reed. At 27 years old and a mother of one child, Cathryn, she was asked by one of the Reading Airshow sponsors if she would attempt to beat the world altitude record set by Grace Huntington. Although she reached and surpassed 26,000 on her aircraft's altimeter, she was faced with the decision of continuing the flight or switching oxygen bottles, but felt the risk of potentially passing out while transferring to another bottle was too great. The decision proved to be the correct one in that she broke the record that was achieved nine years earlier. (Courtesy of Oklahoma University History of Science Collections.)

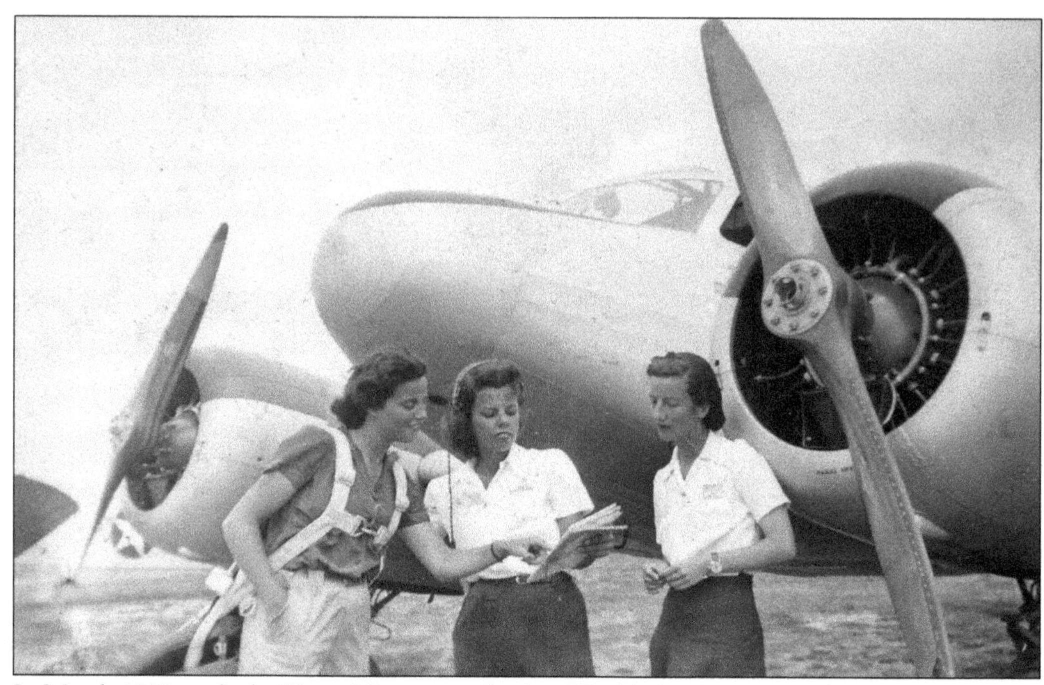

In March 1942, with the United States involved in World War II, two Reading women became heavily involved in promoting aviation to fellow women pilots by organizing a women's flying club at Madeira Field. After purchasing their own Taylorcraft and getting 10 members, Ellen Gery (left) and Emily Hiester (right) were involved in the local Civil Air Patrol squadron and eventually joined the Women Airforce Service Pilots (WASP). (Courtesy of the Women's Collection—Texas Women's University.)

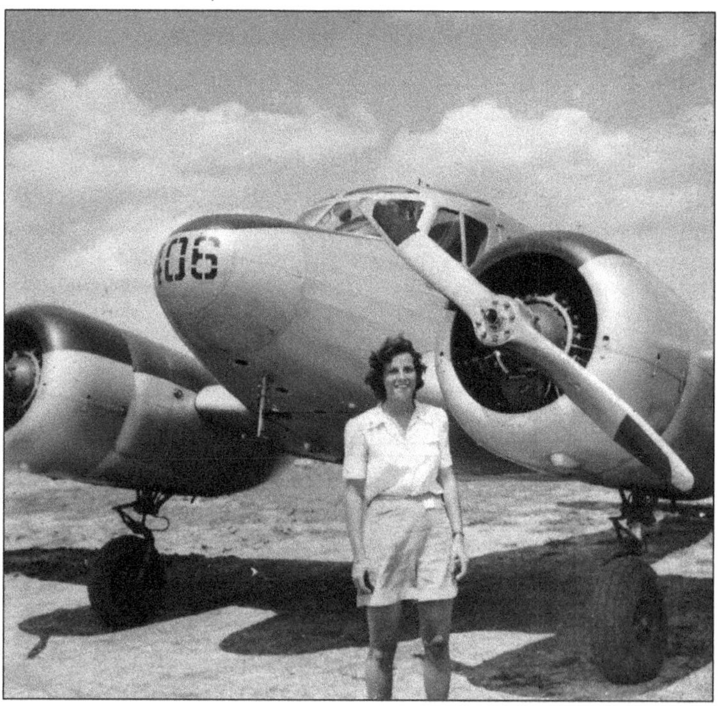

Ellen Gery, pictured in 1943 at Avenger Field in Sweetwater, Texas, supported our country through the Women Airforce Service Pilots, a paramilitary aviation organization of female pilots employed to fly military aircraft under the direction of the US Army Air Forces. In May 1943, Gery graduated among 43 women pilots from the Army Air Forces flying school. She eventually became a school teacher, professional golfer, and real estate businesswoman. (Courtesy of the Women's Collection—Texas Women's University.)

The first pilot's license issued by the state of Pennsylvania was granted in 1928 to Reading resident Harry Bitterman. Seeing a need for aviation in Reading before many others recognized the positive impact it would have on the community and surrounding industries, he, along with four other aviators, formed the Reading Aero Club. Twelve years before the Reading Municipal Airport was built, the club advocated building an airport that met many practical needs. (Courtesy of the Reading Municipal Airport.)

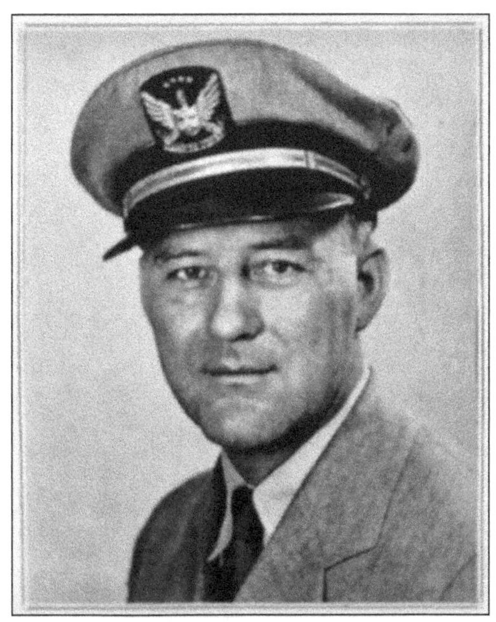

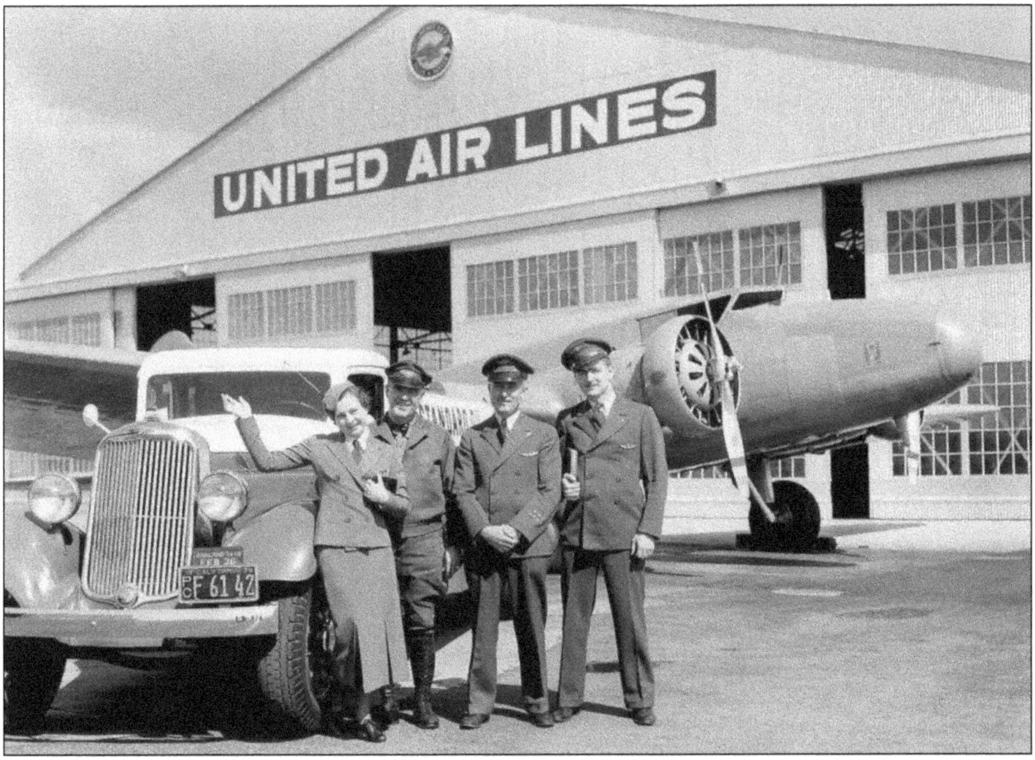

Harry Bitterman, with another Reading aviator as copilot, witnessed Reading's modern airport by representing United Airlines at Reading Municipal Airport's dedication ceremony flying in a 21-passenger Mainliner similar to the one pictured. As captain for United, he achieved numerous company accomplishments including being the first commercial airline pilot to land at Lehigh Valley International Airport in 1935 and also twice breaking United's transport speed records flying the Boeing 247 between Chicago and Cleveland. (Courtesy of United Airlines Historical Foundation.)

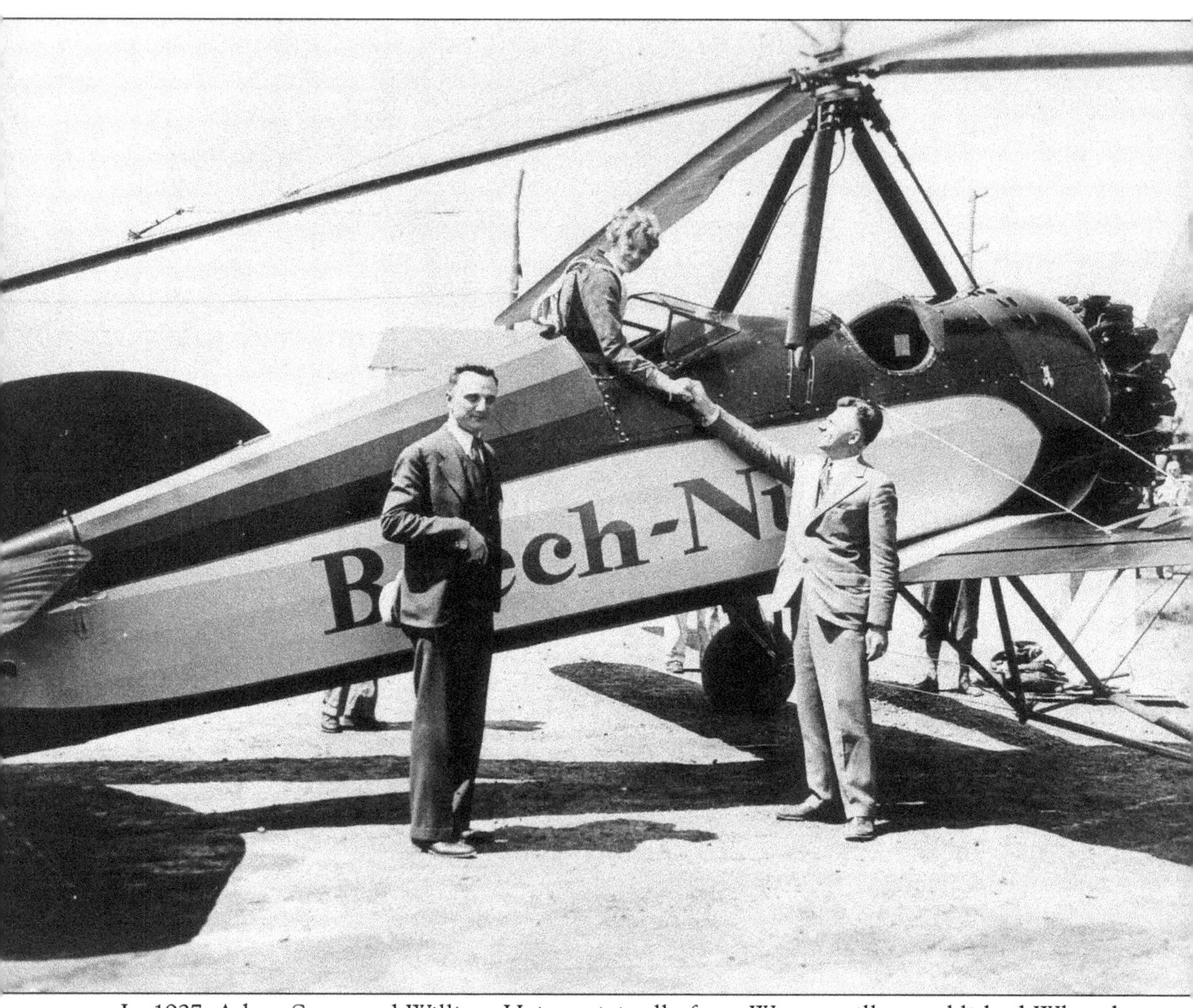

In 1927, Adam Spatz and William Hain, originally from Wernersville, established Whander Field, which was intended to be Reading's first municipal airport. They were among the first Berks Countians to hold commercial pilot licenses. Both contributed greatly to Berks County aviation by forming Reading Airways—the first company in Pennsylvania to receive a charter for commercial flying, making Whander Field the first commercial airport in Reading. In 1932, Spatz was credited with setting a high-altitude record when he went to 11,000 feet without protective clothing. Hain and Spatz were frequent hosts to aviation pioneers such as Maj. Carl Spaatz and barnstormer Clyde Pangborn. This photograph, taken in May 1931, shows Amelia Earhart greeted by William Hain (left) and Adam Spatz when she dropped in at Whander Field in her Beech-Nut Autogiro from Newark, New Jersey. She dined with the men at Indian Queen Manor in Leesport before continuing her transcontinental flight. Before leaving Whander Field, she stated to a large crowd, "I'll try to spend a day in Reading next time." (Courtesy of Terry Sroka.)

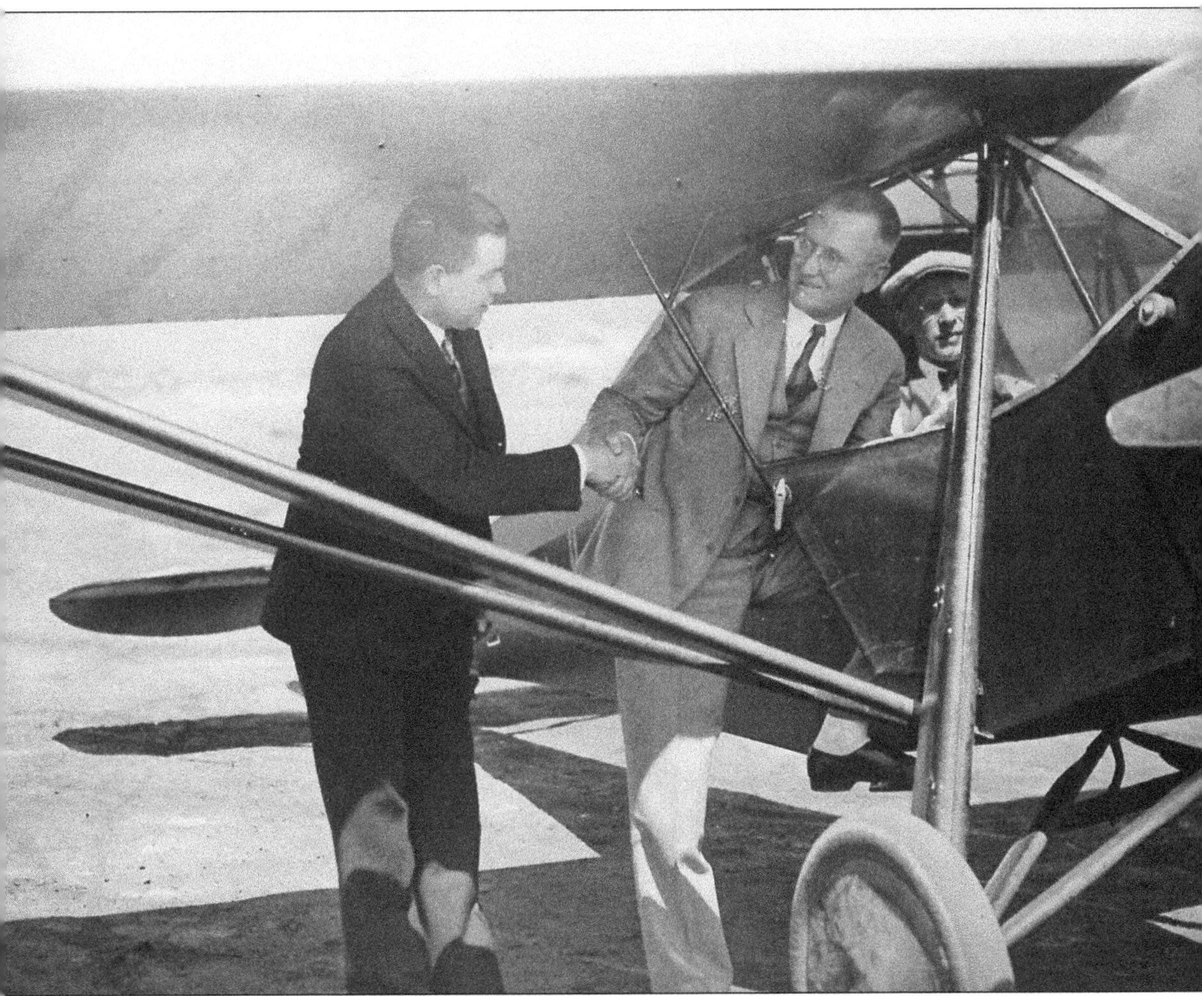

Carl Sisk was a local aviator who promoted aviation in Berks County. In June 1927, Sisk wrote a letter to the mayor stating the importance of building an adequate airport in Reading. His position as financial executive of the former Pomeroy's department store put Reading on navigational maps as a federal airway location with the acquisition of two navigation beacons on Pomeroy buildings. These beacons, located atop the Harrisburg and Reading Pomeroy stores with a two million–candlepower light visible for 40 miles, guided pilots along a route from Pittsburgh to New York by way of Whander Field. By 1939, the Reading beacon was removed and given to the Reading Municipal Airport as a gift from the store's president, George S. Pomeroy (left), shown shaking hands with Carl Sisk in May 1930 before departing to Chicago in a Monocoupe flown by Miles Erbor, one of the first pilots hired by Reading Airways. Sisk became the first person to land at the Reading Municipal Airport with two other passengers—his wire-haired terrier and William Hain—before the field was officially completed. (Courtesy of the Berks History Center.)

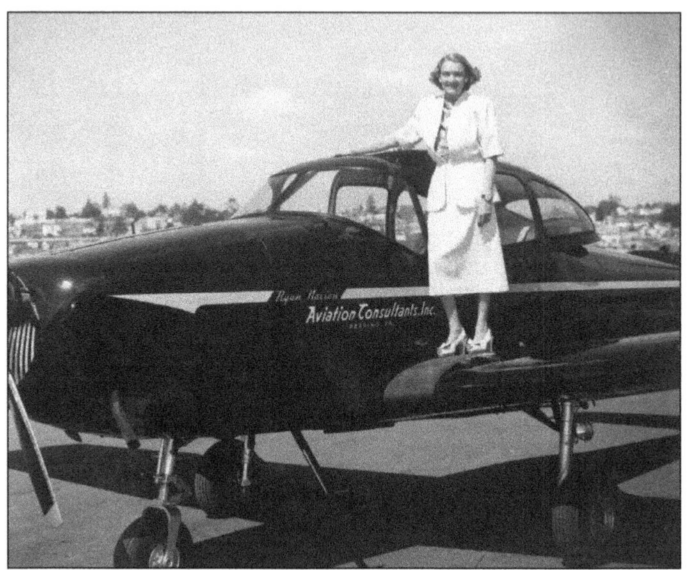

Frances W. Nolde was the wife of a leading Reading industrialist and is shown standing on her crimson Ryan Navion in 1948. She was a fearful flyer but eventually became a prominent aviatrix in Berks County. She said that "air travel was a 'coming thing' and unless she conquered fear, the day would dawn when she'd be sitting home wringing her hands while others were all up in the air." (Courtesy of the San Diego Air & Space Museum.)

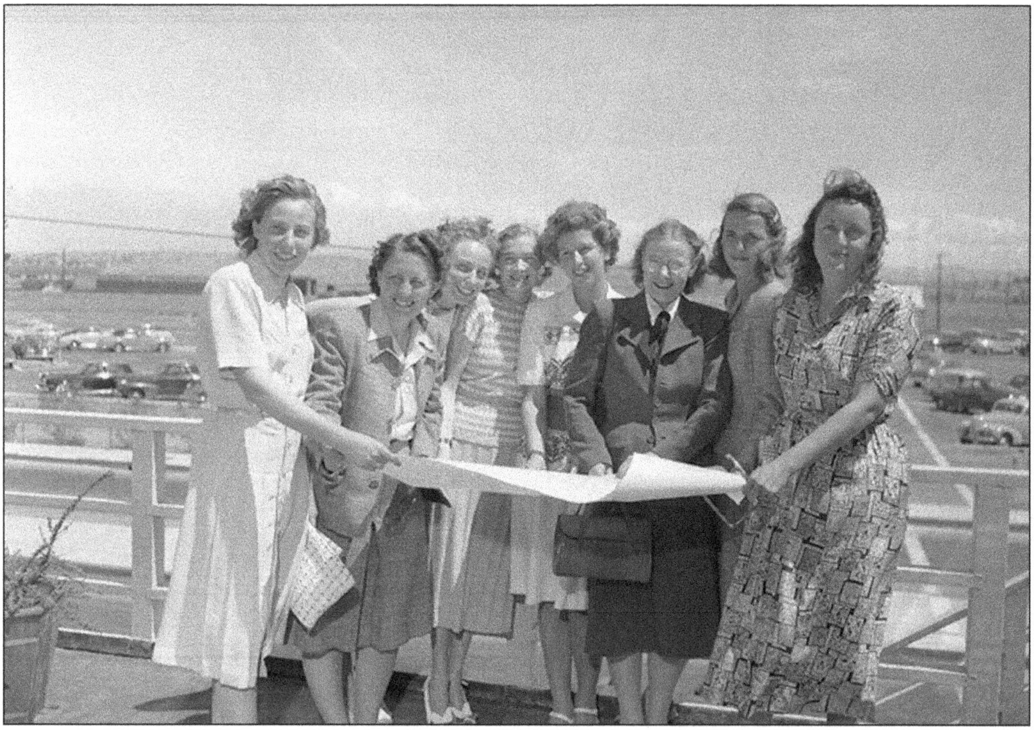

Frances Nolde acquired her pilot's license in 1941 and commercial pilot's license before participating in international air races for women. She won the inaugural transcontinental trophy after successfully beating 16 women, flying 2,450 miles between Palm Springs and Miami. She is shown (center) at Palm Springs in June 1948. She became the first female member of the Reading Airport Commission, director of the National Aeronautic Association, and chairwoman of the Reading Airshow. (Courtesy of the San Diego Air & Space Museum.)

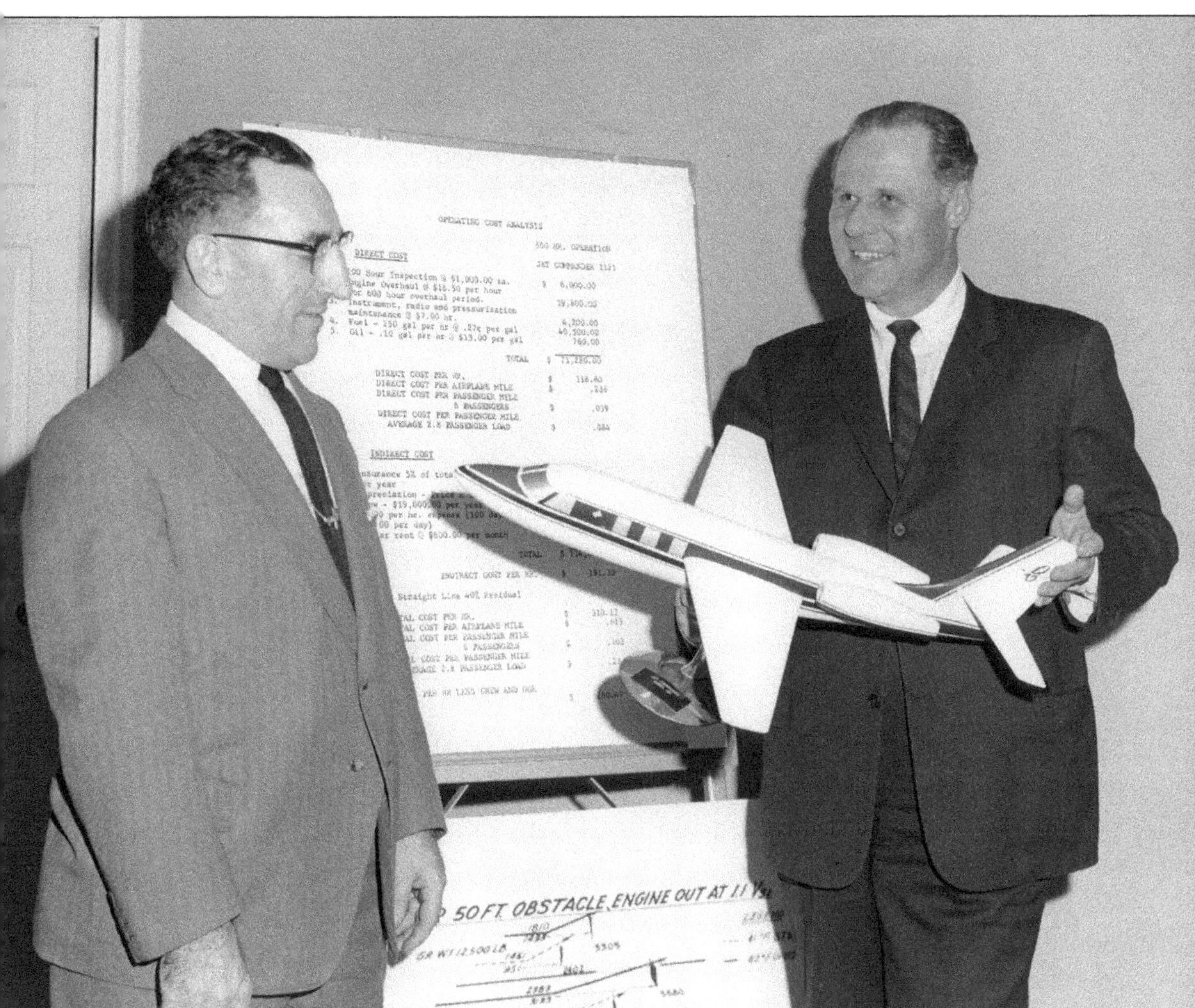

Alvin E. "Gabby" Renninger (left), pictured with Sime Bertolet in front of operational data for the Aero Commander 1121 Jet Commander in 1962, lived in Boyertown and was president of the Aircraft Owners and Pilots Association for 37 years. He was a member of many aviation organizations in the country. He was presented the Veteran of the Year Award in 2000 by the Joint Veteran's Council of Pottstown, Pennsylvania, for organizing fly-overs at Fourth of July parades and flower drops over cemeteries on Memorial Day. In 1966, he was honored by the Aviation Council of Pennsylvania with the Distinguished Aviation Citizen Award, and in 1975, he received the governor's award trophy from the Aviation Council of Pennsylvania for his dedication to the advancement of aviation, giving him the title of "Mr. Aviation of Pennsylvania." (Courtesy of Sime Bertolet.)

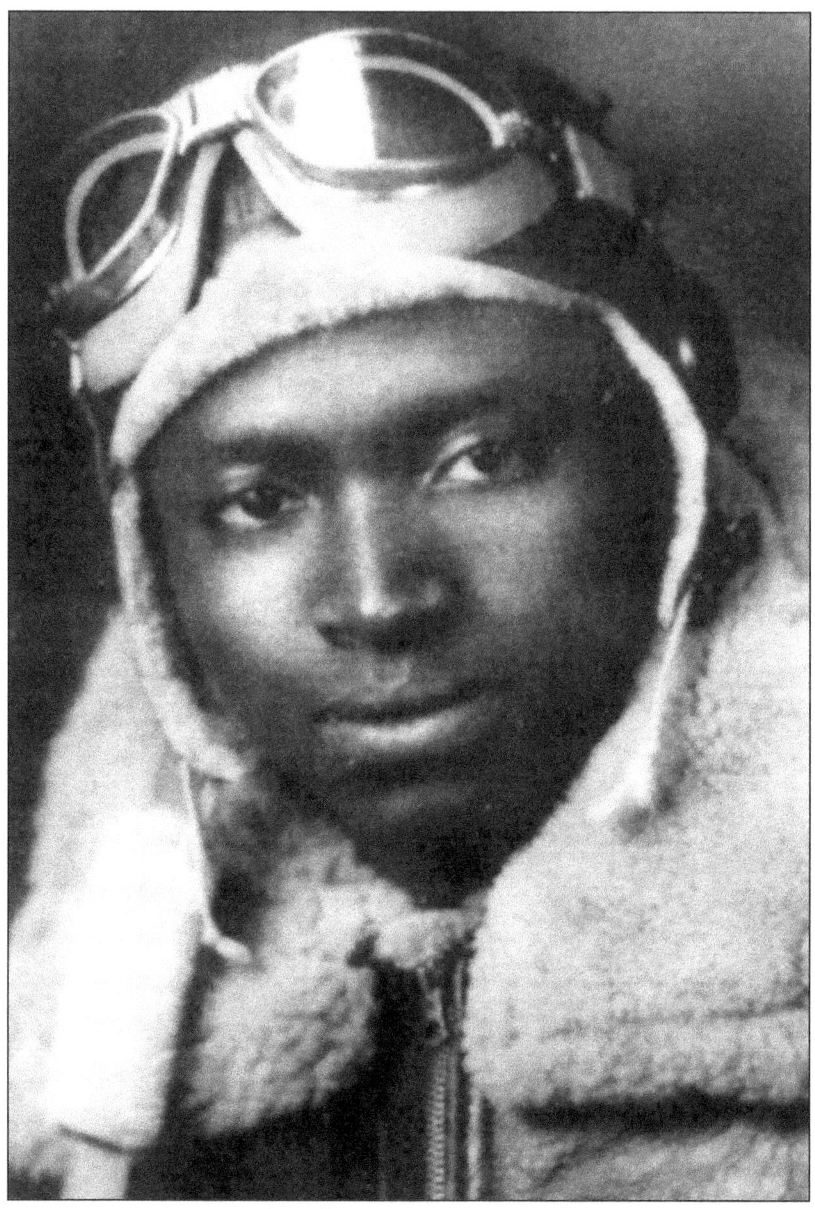

Born in Percy, Mississippi, Roosevelt Miller's curiosity about airplanes was stimulated by a farmer who took him up in his crop-dusting plane. He applied for aviation training with the US Army Air Corps during World War II and became a qualified pilot on May 12, 1943, before enlisting shortly thereafter. As an official member of the Tuskegee Airmen on March 11, 1944, he served overseas with the 99th Fighter Squadron. In February 1946, Miller left the service with several decorations, and because he already had two brothers living in Reading, he settled there and became a member of the local Aircraft Owners and Pilots Association and taught many others to fly at the Reading Airport. In addition to his flying accomplishments and courage as a veteran, he broke racial barriers and graduated from Temple University with a degree in criminal justice before becoming Reading's first black police officer and Reading's first black liquor store employee. Additionally, he founded the George Washington Carver American Legion Post 962 in Reading and operated a barbershop in Reading. (Courtesy of the Central Pennsylvania African American Museum.)

Four

Airfields and Aerial Navigation

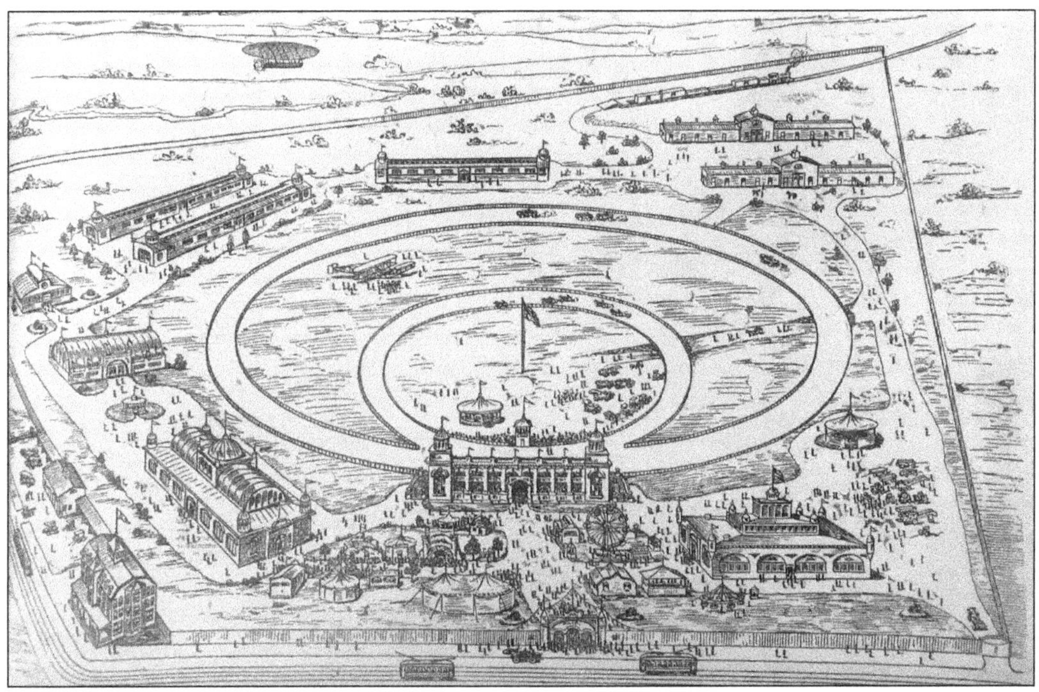

During the early days of Berks County aviation, leaders and aviation pioneers believed Shillington was an ideal location for a track and airport because aviators landed in fields near the county home. Samuel Bigony started the area's first flying operation out of Bigony Field on Lancaster Avenue. Little is known about this field, but it existed until June 1927 when Edwin Lamm, eventual operator of Madeira Field, landed at the field after flying from Texas. (Courtesy of the Berks History Center.)

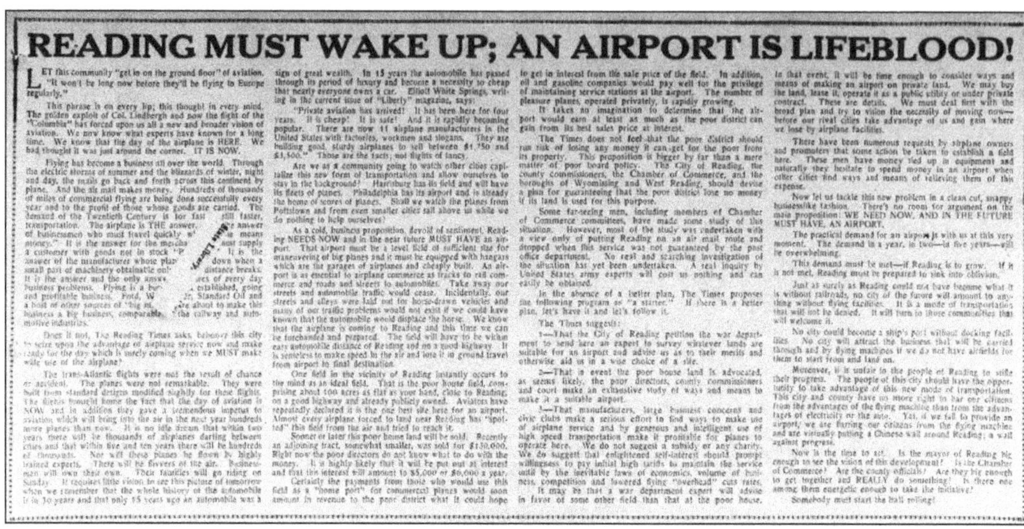

READING MUST WAKE UP; AN AIRPORT IS LIFEBLOOD!

LET this community "get in on the ground floor" of aviation. "It won't be long now before they'll be flying to Europe regularly."

This phrase is on every lip; this thought in every mind. The golden exploit of Col. Lindbergh and now the flight of the "Columbia" has forced upon us all a new and broader vision of aviation. We now know what experts have known for a long time. We know that the day of the airplane is HERE. We had thought it was just around the corner. IT IS NOW.

Flying has become a business all over the world. Through the electric storms of summer and the blizzards of winter, night and day, the mails go back and forth across this continent by plane. And the air mail makes money. Hundreds of thousands of miles of commercial flying are being done successfully every year and to the profit of those whose goods are carried. The demand of the Twentieth Century is for fast ... still faster, transportation. The airplane is THE answer. ... answer of businessmen who must travel quickly w... means money." It is the answer for the mercha... must supply a customer with goods not in stock "P... It is the answer of the manufacturer whose plan... down when a small part of machinery obtainable onl... distance breaks. It is the answer and the only answ... ...ses of every day business problems. Flying is a bu... established, going and profitable business. Ford, W... ...r, Standard Oil and a host of other sources of "big m... are about to make this business a big business, comparable ... the railway and automotive industries.

Does it not, The Reading Times asks, behoove this city to seize upon the advantage of airplane service now and make ready for the day which is surely coming when we MUST make wide use of the airplane?

The trans-Atlantic flights were not the result of chance or accident. The planes were not remarkable. They were built from standard designs modified slightly for these flights. The flights brought home the fact that the day of aviation is NOW and in addition they gave a tremendous impetus to aviation which will bring into the air in the next year hundreds more planes than now. It is no idle dream that within two years there will be thousands of airplanes darting between cities and that within five and ten years there will be hundreds of thousands. Nor will these planes be flown by highly trained experts. There will be flivvers of the air. Businessmen will own their own. Their families will go riding on Sunday. It requires little vision to see this picture of tomorrow when we remember that the whole history of the automobile is in 30 years and that only 15 years ago an automobile was a sign of great wealth. In 15 years the automobile has passed through its period of luxury and become a necessity so cheap that nearly everyone owns a car. Elliott White Springs, writing in the current issue of "Liberty" magazine, says:

"Private aviation has arrived! It has been here for four years. It is cheap! It is safe! And it is rapidly becoming popular. There are now 11 airplane manufacturers in the United States with factories, workmen and slogans. They are building good, sturdy airplanes to sell between $1,750 and $3,500." Those are the facts; not flights of fancy.

Are we as a community going to watch other cities capitalize this new form of transportation and allow ourselves to stay in the background? Harrisburg has its field and will have its fleets of planes. Philadelphia has its airport and is already the home of scores of planes. Shall we watch the planes from Pottstown and from even smaller cities sail above us while we do nothing to help ourselves?

As a cold, business proposition, devoid of sentiment, Reading NEEDS NOW and in the near future MUST HAVE an airport. That airport must be a level field of sufficient size for maneuvering of big planes and it must be equipped with hangars which are the garages of airplanes and cheaply built. An airport is as essential to airplane commerce as tracks to rail commerce and roads and streets to automobiles. Take away our streets and automobile traffic would cease. Incidentally, our streets and alleys were laid out for horse-drawn vehicles and many of our traffic problems would not exist if we could have known that the automobile would displace the horse. We know that the airplane is coming to Reading and this time we can be forehanded and prepared. The field will have to be within easy automobile distance of Reading and on a good highway. It is senseless to make speed in the air and lose it in ground travel from airport to final destination.

One field in the vicinity of Reading instantly occurs to the mind as an ideal field. That is the poor house field, comprising about 100 acres as flat as your hand, close to Reading, on a good highway and already publicly owned. Aviators have repeatedly declared it is the one best site here for an airport. Almost every airplane forced to land near Reading has "spotted" this field from the air and tried to reach it.

Sooner or later this poor house land will be sold. Recently an adjoining tract, somewhat smaller, was sold for $130,000. Right now the poor directors do not know what to do with the money. It is highly likely that it will be put out at interest and that the interest will amount to $5,000 or $6,000 a year.

Certainly the payments from those who would use this field as a "home port" for commercial planes would soon amount in revenue to the poor district what it could hope to get in interest from the sale price of the field. In addition, oil and gasoline companies would pay well for the privilege of maintaining service stations at the airport. The number of pleasure planes, operated privately, is rapidly growing.

It takes no imagination to determine that the airport would earn at least as much as the poor district can gain from its best sales price at interest.

The Times does not feel that the poor district should run risk of losing any money it can get for the poor from its property. This proposition is bigger by far than a mere matter of poor board policy. The City of Reading, the county commissioners, the Chamber of Commerce, and the boroughs of Wyomissing and West Reading, should devise a plan for guaranteeing that the poor district lose no money if its land is used for this purpose.

Some far-seeing men, including members of Chamber of Commerce committees, have made some study of this situation. However, most of the study was undertaken with a view only of putting Reading on an air mail route and dropped when this service was not guaranteed by the post office department. No real and searching investigation of the situation has yet been undertaken. A real inquiry by United States army experts will cost us nothing and can easily be obtained.

In the absence of a better plan, The Times proposes the following program as "a starter." If there is a better plan, let's have it and let's follow it.

The Times suggests:

1—That the City of Reading petition the war department to send here an expert to survey whatever lands are suitable for an airport and advise us as to their merits and otherwise aid us in a wise choice of a site.

2—That in event the poor house land is advocated, as seems likely, the poor directors, county commissioners and court make an exhaustive study of ways and means to make it a suitable airport.

3—That manufacturers, large business concerns and civic clubs make a serious effort to find ways to make use of airplane service and by generous and intelligent use of high speed transportation make it profitable for planes to operate here. We do not suggest a subsidy or any charity. We do suggest that enlightened self-interest should prompt willingness to pay initial high tariffs to maintain the service until by the inevitable laws of economics, volume of business, competition and lowered flying "overhead" cuts rates.

It may be that a war department expert will advise in favor of some other field than that at the poor house. In that event, it will be time enough to consider ways and means of making an airport on private land. We may buy the land, lease it, operate it as a public utility or under private contract. These are details. We must deal first with the broad plan and try to vision the necessity of moving now—before our rival cities take advantage of us and gain where we lose by airplane facilities.

There have been numerous requests by airplane owners and promoters that some action be taken to establish a field here. These men have money tied up in equipment and naturally they hesitate to spend money in an airport when other cities find ways and means of relieving them of this expense.

Now let us tackle this new problem in a clean cut, snappy businesslike fashion. There's no room for argument on the main proposition: WE NEED NOW, AND IN THE FUTURE MUST HAVE, AN AIRPORT.

The practical demand for an airport is with us at this very moment. The demand in a year, in two—in five years—will be overwhelming.

This demand must be met—if Reading is to grow. If it is not met, Reading must be prepared to sink into oblivion.

Just as surely as Reading could not have become what it is without railroads, no city of the future will amount to anything without flying facilities. It is a mode of transportation that will not be denied. It will turn to those communities that will welcome it.

No city could become a ship's port without docking facilities. No city will attract the business that will be carried through and by flying machines if we do not have airfields for them to start from and land on.

Moreover, it is unfair to the people of Reading to stifle their progress. The people of this city should have the opportunity to take advantage of this new mode of transportation. This city and county have no more right to bar our citizens from the advantages of the flying machine than from the advantages of electricity or the auto. Yet, if we fail to provide an airport, we are barring our citizens from the flying machine and are virtually putting a Chinese wall around Reading; a wall against progress.

Now is the time to act. Is the mayor of Reading big enough to see the vision of this development? Is the Chamber of Commerce? Are the county officials? Are they big enough to get together and REALLY do something? Is there one among them energetic enough to take the initiative?

Somebody must start the ball rolling!

Plans were developed for the city's airport to be in Shillington, but residents blocked its construction. Instead, Whander Field was selected to serve this purpose during the 1920s even though Madeira Field already existed. Eventually, neither Whander Field nor Madeira Field were adequate to accommodate changes in aviation. One month following Charles Lindbergh's transatlantic flight, this June 10, 1927, *Reading Times* article challenged the city to "wake up" and "get on the ground floor of aviation." (Author's collection.)

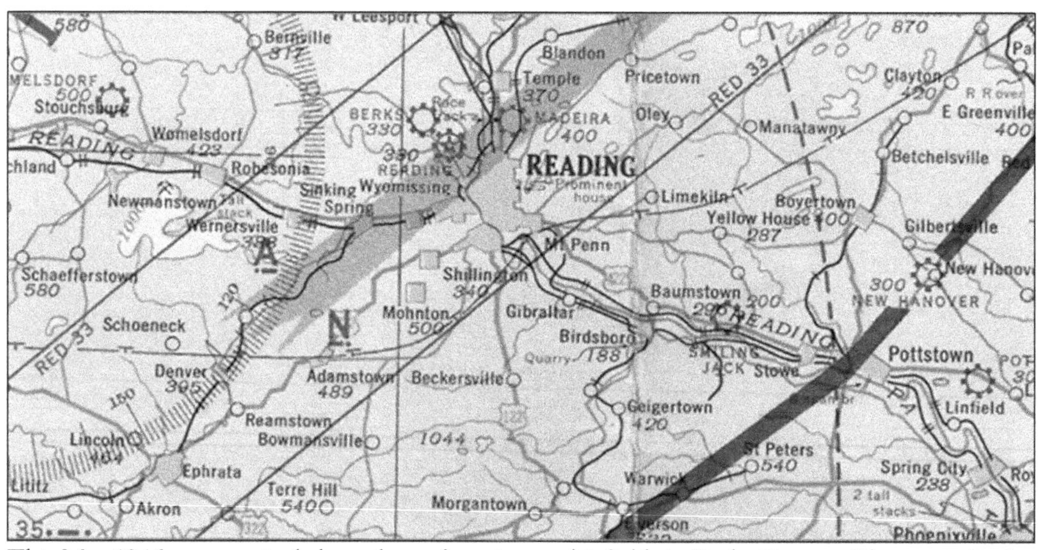

This May 1946 aeronautical chart shows five airports/airfields in Berks County. They were Berks, Womelsdorf, Smiling Jack, Madeira, and Reading. Whander did not appear because it had been closed when the chart was published. Whander closed in the late 1920s, reopened in 1938, and reopened again for a short duration in July 1946 before being permanently closed. (Author's collection.)

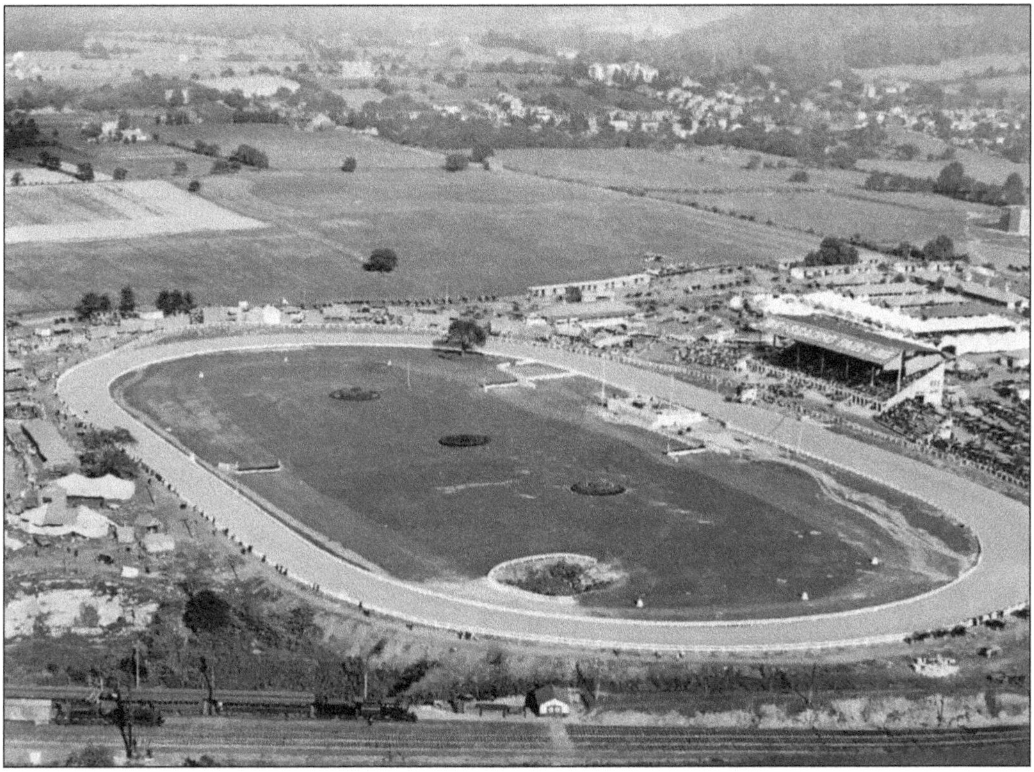

Madeira Field was located east of the Fifth Street Highway, across from the Reading Fairgrounds now occupied by the Madeira shopping center. The picture above shows a biplane to the right of the field across from the raceway. Originally, the Madeira family owned the property until Charles Madeira, a farmer and aviation proponent who once hoped his field would become the city's airport, filed an application for an airport license in 1919 and modified the farm for airport use by converting his barn to a hangar. The barn eventually housed nine airplanes before other structures were added to the field. Once considered the oldest Berks County airport in operation, Madeira Field, like many other airports in the United States, began with barnstormers using the field as a place for aerial circuses and providing passenger flights for a fee. These images are from volume 24 of *The Passing Scene*. (Both, courtesy of George M. Meiser IX.)

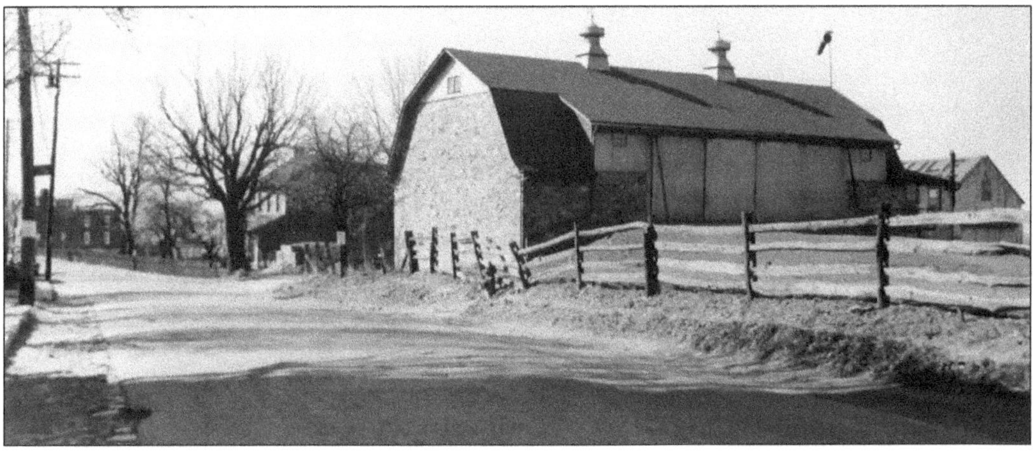

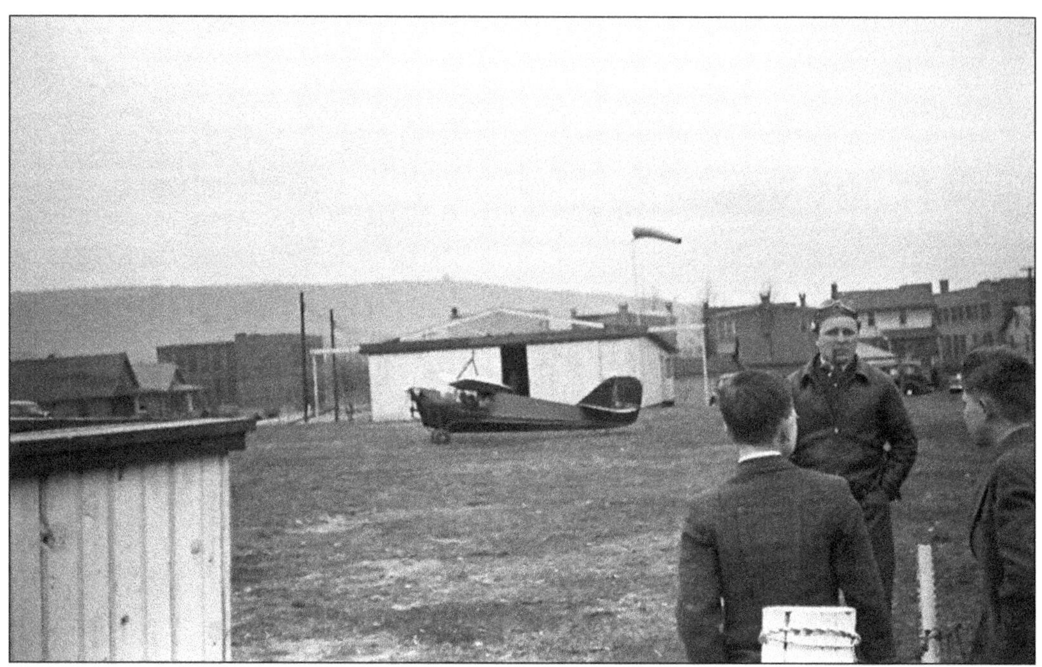

Above, looking southeast, an Aeronca sits waiting for flight training use as Eddie Nibur, with his ever-present pipe, talks with two boys. Below, looking northeast, some people are departing a Stinson, used for commercial purposes, as cars sit along the perimeter of the grass runway. Madeira Field had two runways, but when Eddie Nibur assumed operations at Madeira, the runways were lengthened and reconfigured, with the east to west runway measuring 2,300 feet and another runway from northwest to southeast measuring 2,000 feet. The airport was officially dedicated on April 2, 1934, with the years of 1934 to 1940 seeing the greatest growth as a first-class airport. (Both, courtesy of Ron Ciervo.)

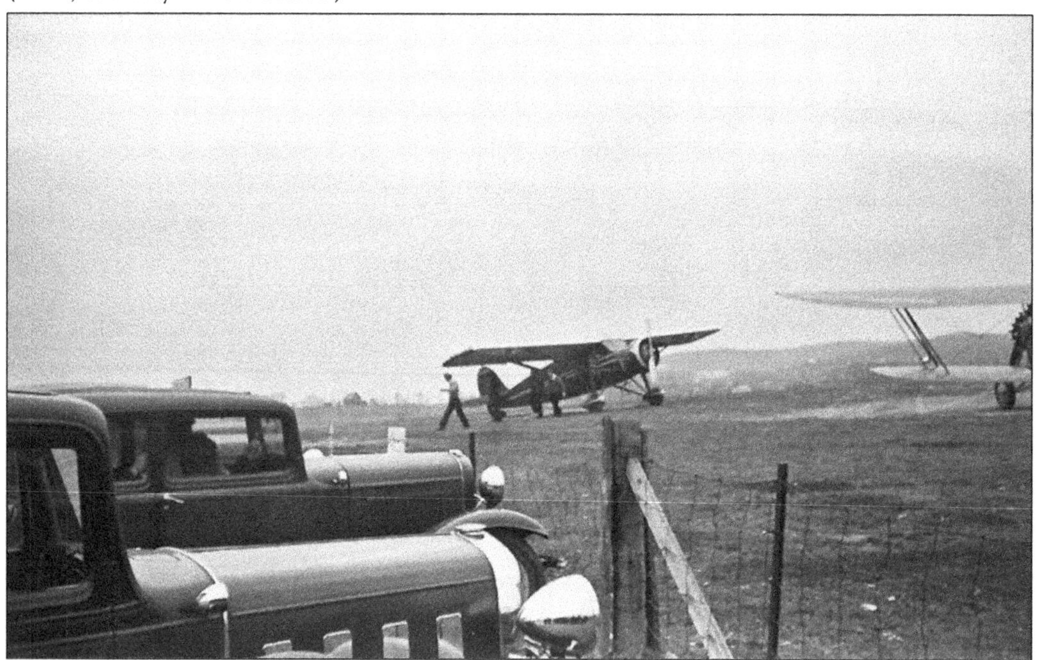

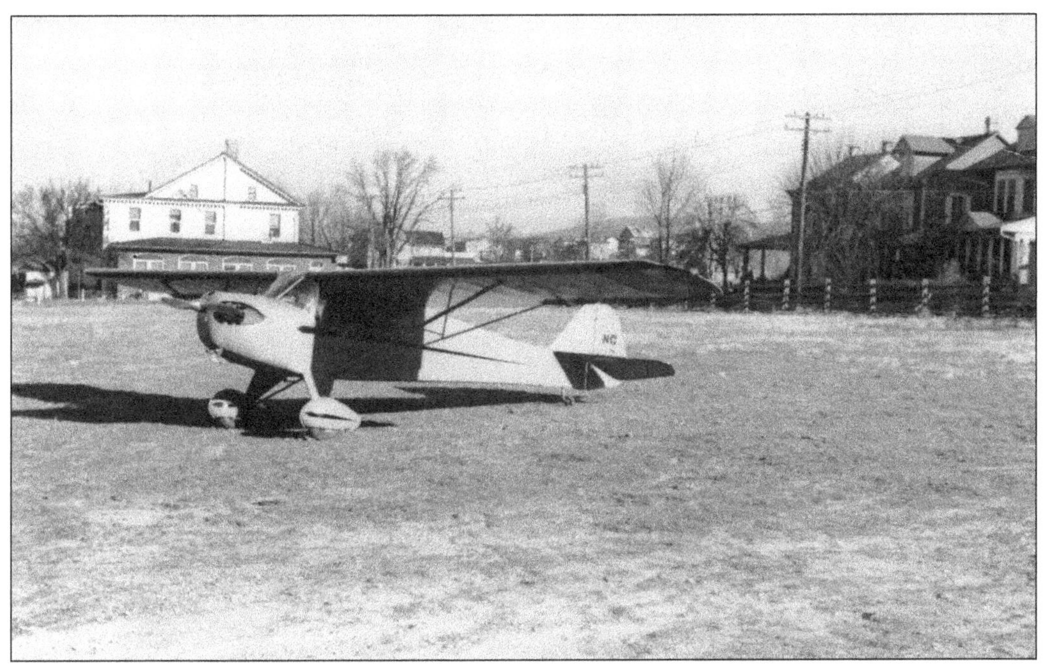

Above, a Taylorcraft Model A, registration No. NC20302, is parked on the south side of the Madeira Field during the mid-1930s. The plane was purchased by Eddie Nibur on January 15, 1938, and used for flight training. It was powered by a four-cylinder Continental engine. The image of the Stinson and Aeronca below was featured in a June 1935 *Reading Times* story about Madeira Field's air circus. The caption was "A Hen and Her Chick." The Stinson Airliner was at the field only as a feature attraction during the air carnival event, but the Taylorcraft was a permanent resident at Madeira. The Taylorcraft two-cylinder had a top speed of 95 miles per hour and used 15 gallons of fuel per hour. (Both, courtesy of Paul Nibur.)

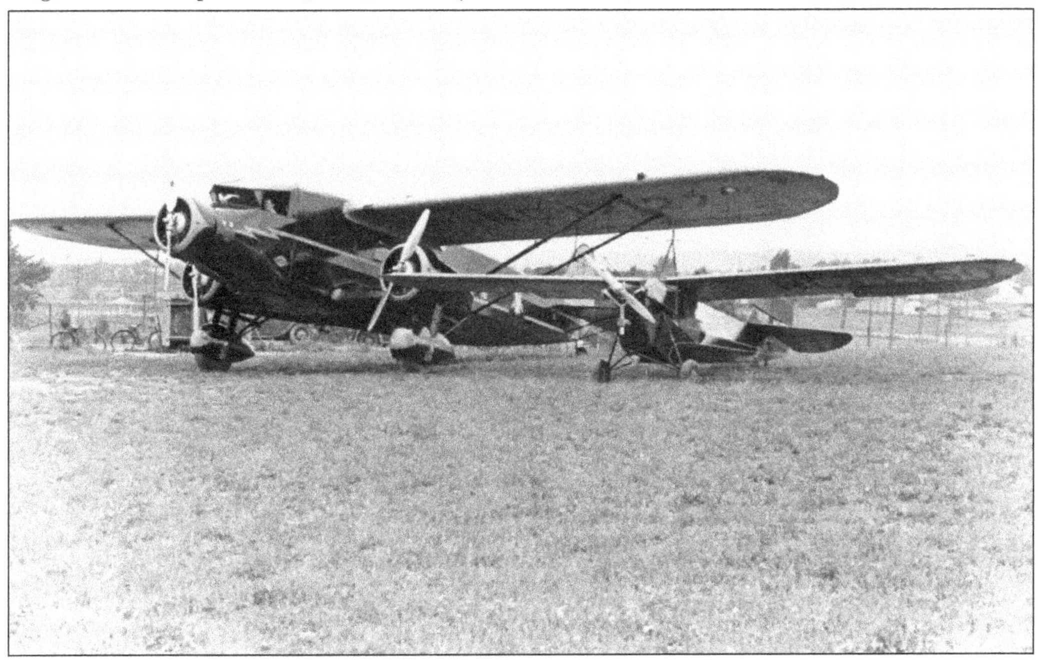

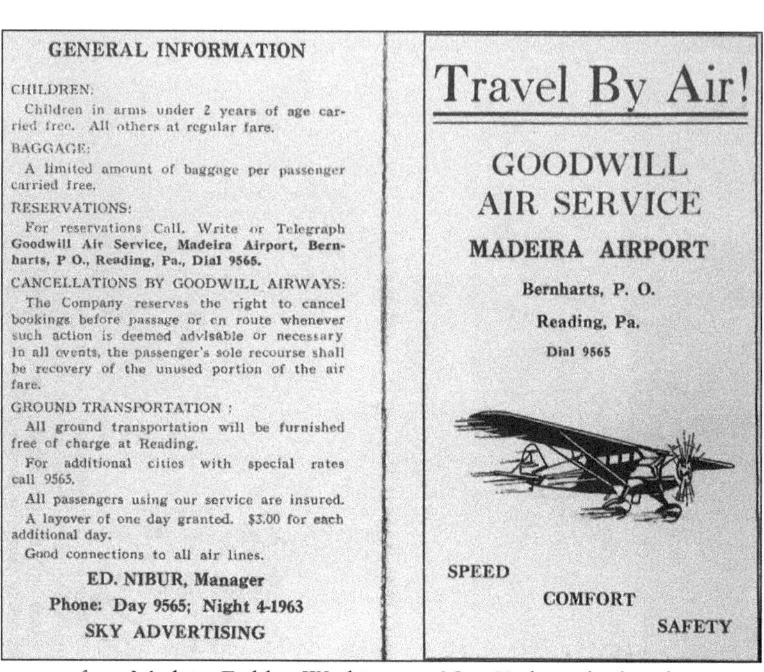

GENERAL INFORMATION

CHILDREN:

Children in arms under 2 years of age carried free. All others at regular fare.

BAGGAGE:

A limited amount of baggage per passenger carried free.

RESERVATIONS:

For reservations Call, Write or Telegraph **Goodwill Air Service, Madeira Airport, Bernharts, P O., Reading, Pa., Dial 9565.**

CANCELLATIONS BY GOODWILL AIRWAYS:

The Company reserves the right to cancel bookings before passage or en route whenever such action is deemed advisable or necessary In all events, the passenger's sole recourse shall be recovery of the unused portion of the air fare.

GROUND TRANSPORTATION :

All ground transportation will be furnished free of charge at Reading.

For additional cities with special rates call 9565.

All passengers using our service are insured.

A layover of one day granted. $3.00 for each additional day.

Good connections to all air lines.

ED. NIBUR, Manager

Phone: Day 9565; Night 4-1963

SKY ADVERTISING

Travel By Air!

GOODWILL
AIR SERVICE

MADEIRA AIRPORT

Bernharts, P. O.

Reading, Pa.

Dial 9565

SPEED

COMFORT

SAFETY

Charter flights went from Madeira Field to Washington, New York, and other destinations. Maryland flights were successful because many Berks Countians loved to bet on horses. Supplemental income came from aerial advertisements, especially for Reading Electric Company and Philadelphia Electric Company. Flights were added to Boston and Atlantic City between 1937 and 1938, and the airport saw increased flight instruction. July 1936 brought the first air races in local history, one of only eight across the country that year. That same month, Eddie Nibur conducted a dramatic air show with acrobatics dropping for the first time a woman parachutist from 9,000 feet above the airfield. Traffic along the Fifth Street Highway was jammed as onlookers watched by the novel aerial feats. (Both, courtesy of Terry Sroka.)

FOR POINTS NOT LISTED BELOW CALL GOODWILL AIR SERVICE, 9565

CONNECTIONS WITH ALL AIR LINES

Destination	Mileage	1 Pass.	2 Pass.	3 Pass.	4 Pass.
Atlantic City	100	$ 26.00	$ 14.00	$ 11.00	$ 9.00
Allentown	30	8.00	4.25	3.25	2.75
Albany	200	52.00	28.00	22.00	18.00
Altoona	130	34.00	18.50	14.00	12.00
Atlanta	660	171.60	92.50	70.40	59.40
Baltimore	85	22.00	12.00	9.25	7.75
Boston	290	75.50	40.60	31.00	26.50
Buffalo	230	59.80	32.25	24.50	21.00
Chicago	625	162.50	87.50	66.50	56.50
Cleveland	315	82.00	45.00	34.00	29.00
Detroit	410	106.60	57.50	43.75	39.25
Harrisburg	50	13.00	7.00	5.50	4.50
Miami	1105	287.50	154.75	117.50	100.00
Montreal	400	104.00	56.00	44.00	36.00
New York City	100	26.00	14.00	11.00	9.00
Philadelphia	50	13.00	7.00	5.50	4.50
Pittsburgh	210	54.60	29.50	22.50	19.00
Syracuse	185	48.00	25.90	19.90	16.75
Toledo	410	106.60	57.50	43.75	39.25
Trenton	65	16.60	9.25	7.00	6.00
Washington	115	30.00	17.00	15.50	10.50
Wilkes-Barre	65	16.00	9.25	7.00	6.00

ALL PRICES ARE ROUND TRIP FARES PER PASSENGER

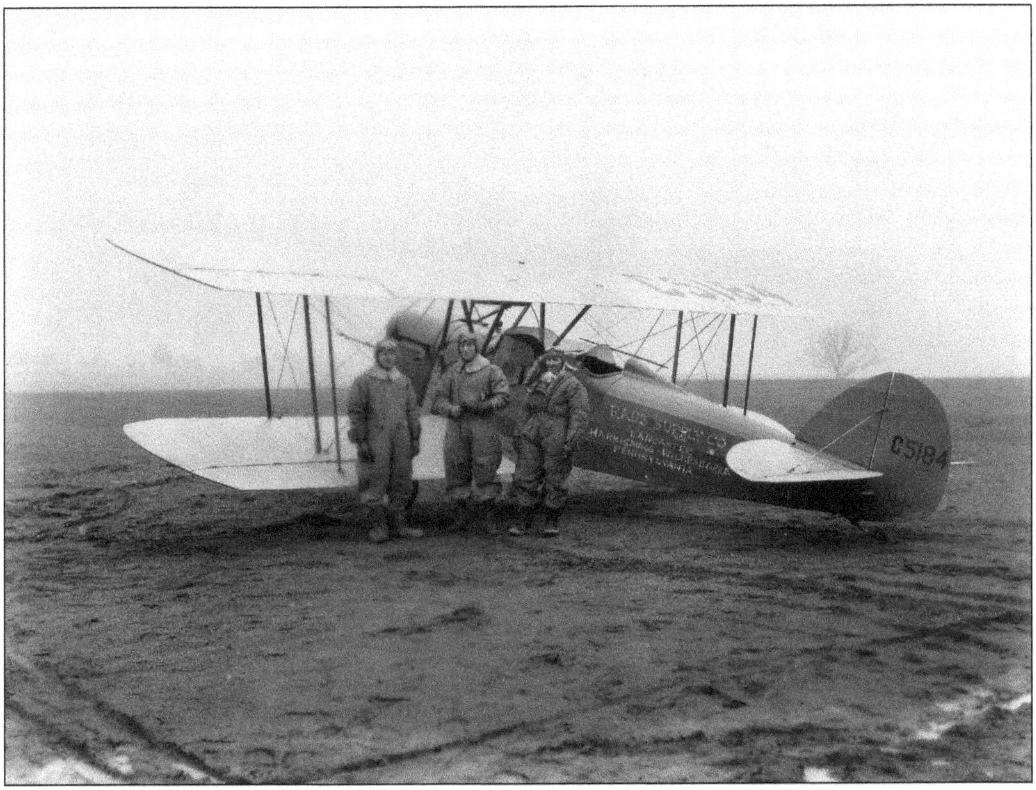

In 1934, three pilots made the first delivery of plumbing supplies from Raub Supply Company of Lancaster to Reading's Madeira Field. Many of the plumbers were city plumbing officials, plumbing and heating contractors, and members of the Master Plumber's Association. To the right is Raub's advertisement commemorating the event, which is very commonplace today. The advertisement shows the pilots loading the plane and notes the names of those who received the shipment. (Both, courtesy of the Lancaster Historical Society.)

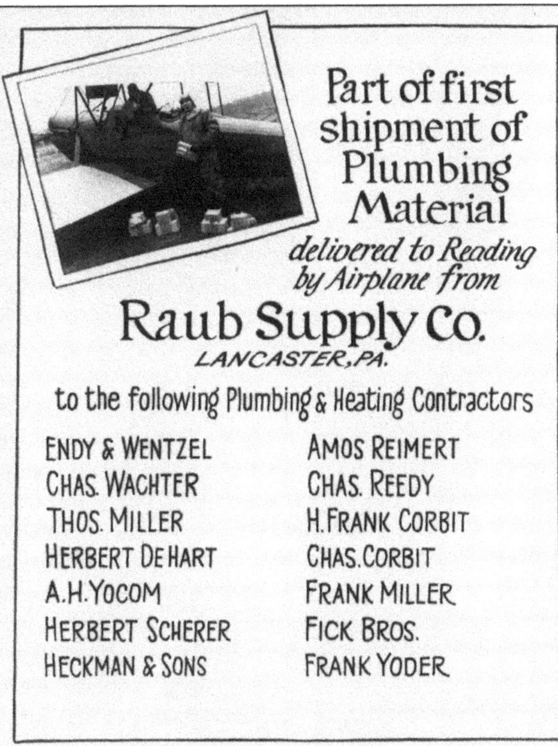

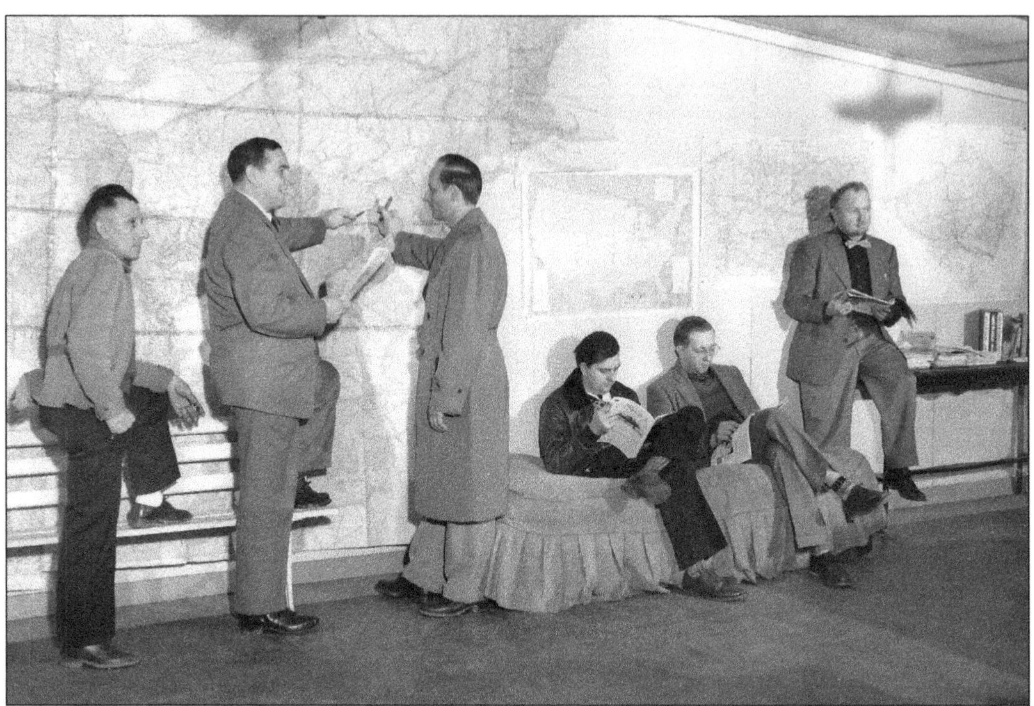

The Reading Aero Club is one of the oldest flying organizations in the country. It relocated to Madeira Field in 1935 after formally beginning at Whander Field in 1932. The club, with members seen above relaxing in the 1950s at Berks Airport, moved to the Reading Municipal Airport in 1965. It still exists; however, it had to eventually relocate to Berks Airport in 1938 because Madeira Field was too small and its hangars were less than ideal for meetings. In February 1940, another club, the Keystone Aero Club, was organized at Madeira Field to fill the void of camaraderie and aviation knowledge shared by the Reading Aero Club. (Both, courtesy of the Reading Aero Club.)

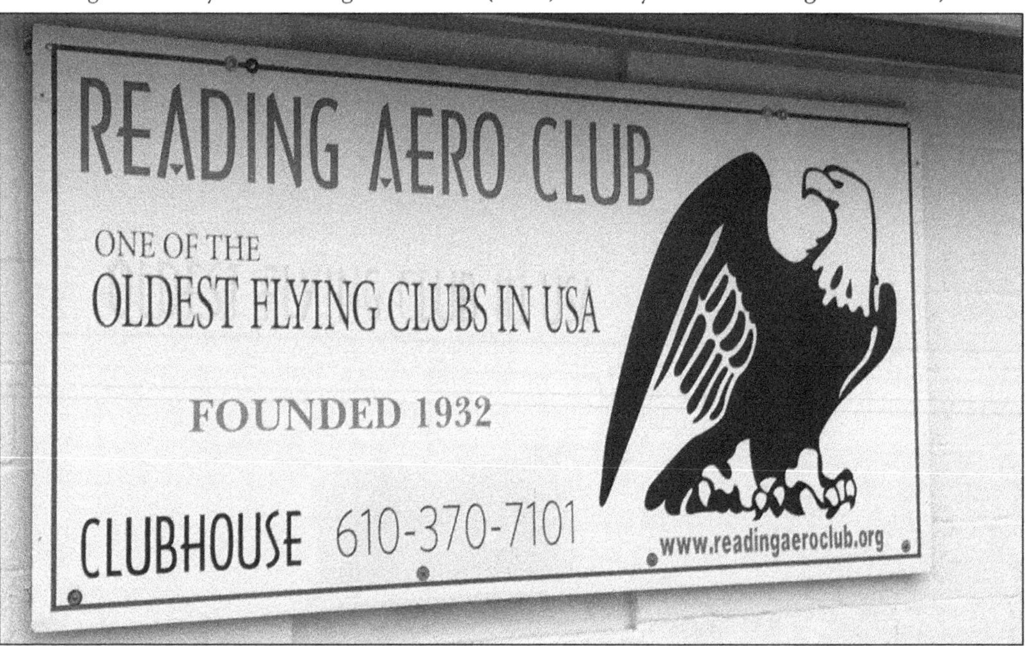

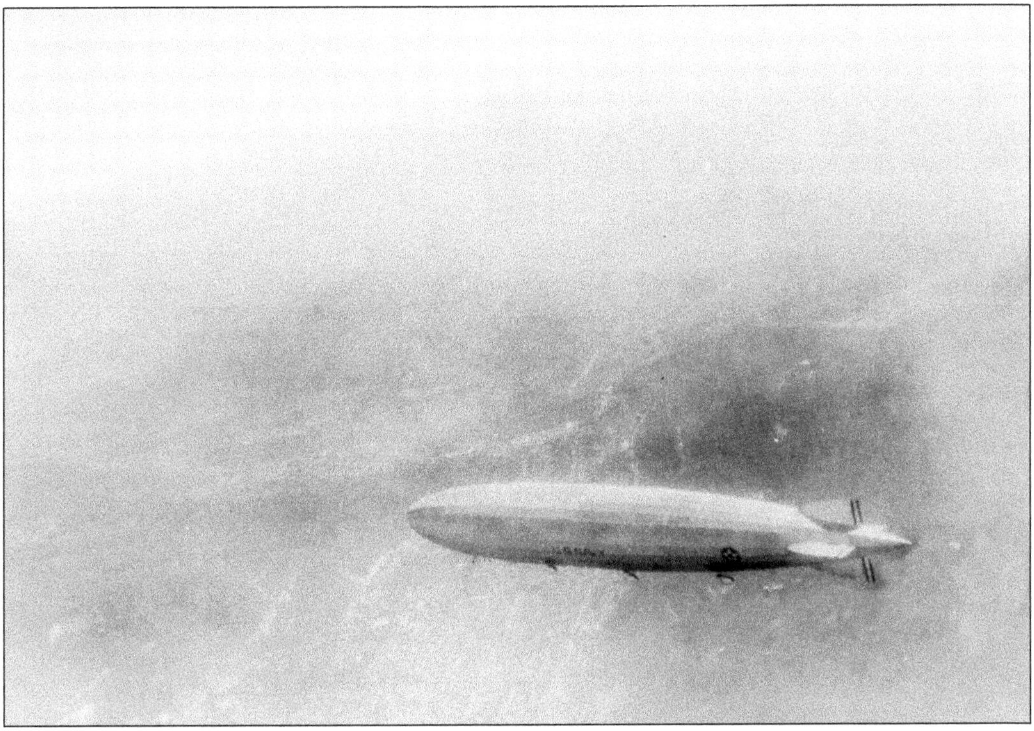

One of the first known dirigibles to fly over Berks County was the USS *Los Angeles*, the most successful of the US Navy's rigid airships, manufactured in Germany from 1922 to 1924 and decommissioned in 1932. The photograph above was taken by another aircraft flying over Reading in marginal weather. At right, the zeppelin can be seen to the northeast over the Washington Street area in August 1929. It was said that some influential citizens were able to communicate with the dirigible's commander, requesting it fly over Reading during its around-the-world voyage. The image at right is from volume 23 of *The Passing Scene*. (Above, courtesy of Terry Sroka; right, courtesy of George M. Meiser IX.)

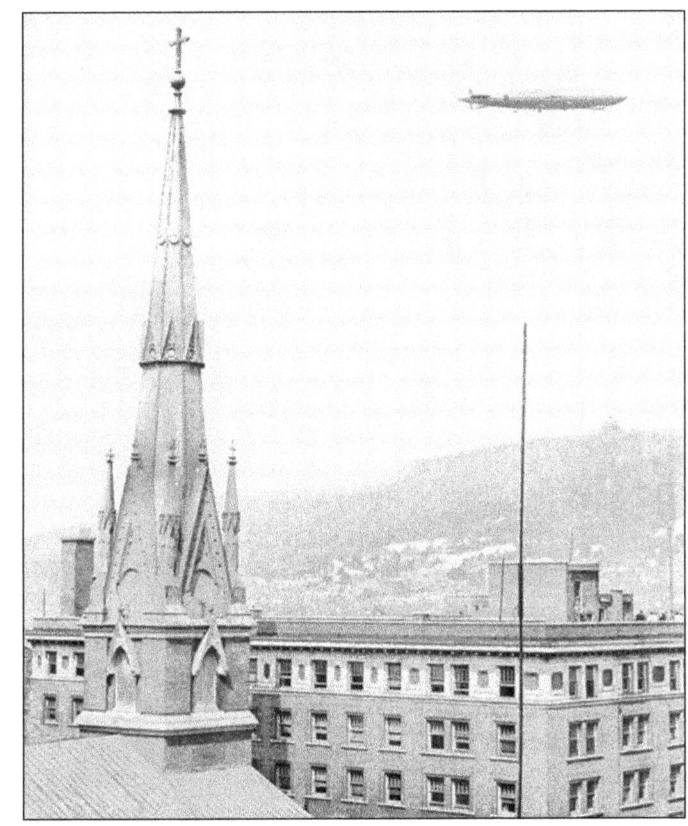

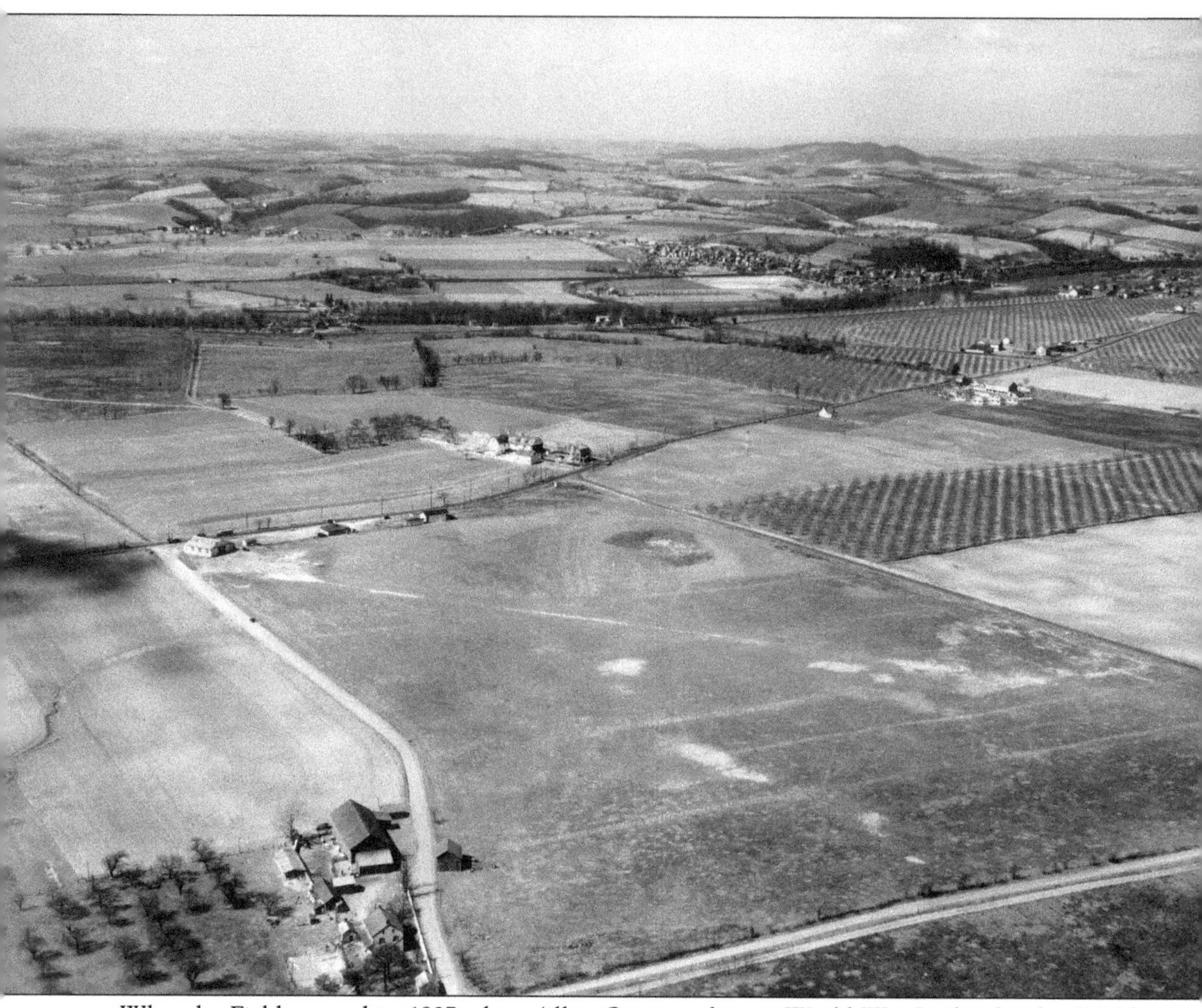

Whander Field opened in 1927 when Allen Owen, a former World War I pilot for the British Air Force, landed his plane in a Leesport field. At that time, he surveyed the Reading area and, with support from John P. Wolfinger along with other aviators and businessmen, Whander Field became a reality. The field's name, derived from the first letter of each founder's last name, was scheduled to be the city of Reading's airport and was said to have everything an airport needed. Since Lake Ontelaunee was close to the airport, consideration was given to having water landing facilities as part of the airport, but the idea was later scrubbed. The field presented challenges to pilots because it was on rising terrain. To the west along the Pottsville Pike were telephone and electric wires and to the east were trees and the lake. Whander was regarded as the county's first commercial airport, allowing transportation of passengers, freight, and airmail. It was managed by Reading Airways. (Courtesy of the Hagley Museum and Library.)

POSTPONED

Owing to the past four days of rain the opening of WHANDER Field, Reading's new commercial airport, has been postponed one week, to Oct. 29, 1927.

A program of events will be announced later.

READING AIRWAYS

Although Whander Field's dedication was delayed because of four days of inclement weather, the following week brought an elaborate air program that drew more than 10,000 individuals to witness what was advertised as "Pennsylvania's biggest air show." The Army, Navy, and Marines sent a multitude of planes to the event. The Marines performed an aerial acrobatics show for the crowd, and Berks County received its first airmail. The mail originated from Des Moines, Iowa, and was flown from Philadelphia's Pitcarin Field to Whander Field. (Above, author's collection; right, courtesy of Sime Bertolet.)

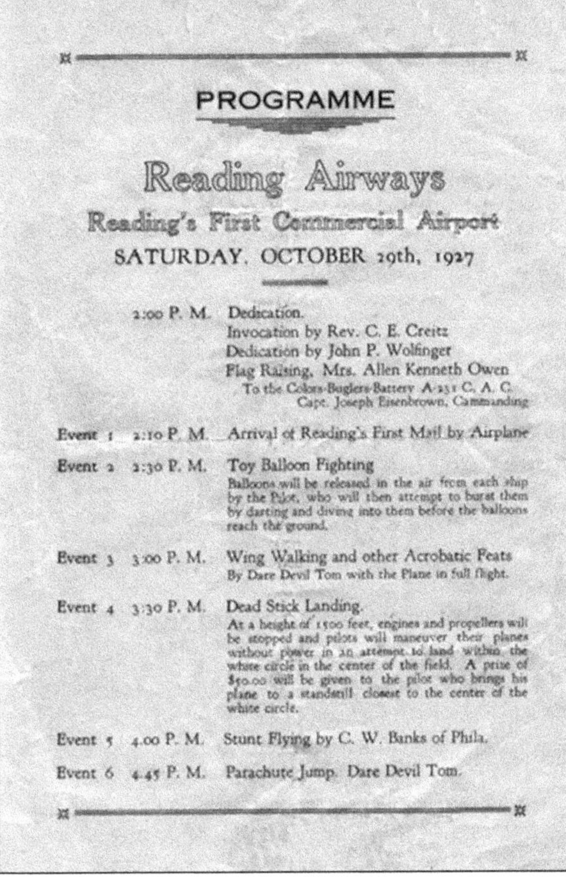

PROGRAMME

Reading Airways

Reading's First Commercial Airport

SATURDAY, OCTOBER 29th, 1927

	2:00 P. M.	Dedication. Invocation by Rev. C. E. Creitz Dedication by John P. Wolfinger Flag Raising, Mrs. Allen Kenneth Owen To the Colors-Buglers-Battery A-231 C. A. C. Capt. Joseph Eisenbrown, Commanding
Event 1	2:10 P. M.	Arrival of Reading's First Mail by Airplane
Event 2	2:30 P. M.	Toy Balloon Fighting Balloons will be released in the air from each ship by the Pilot, who will then attempt to burst them by darting and diving into them before the balloons reach the ground.
Event 3	3:00 P. M.	Wing Walking and other Acrobatic Feats By Dare Devil Tom with the Plane in full flight.
Event 4	3:30 P. M.	Dead Stick Landing. At a height of 1500 feet, engines and propellers will be stopped and pilots will maneuver their planes without power in an attempt to land within the white circle in the center of the field. A prize of $50.00 will be given to the pilot who brings his plane to a standstill closest to the center of the white circle.
Event 5	4.00 P. M.	Stunt Flying by C. W. Banks of Phila.
Event 6	4.45 P. M.	Parachute Jump. Dare Devil Tom.

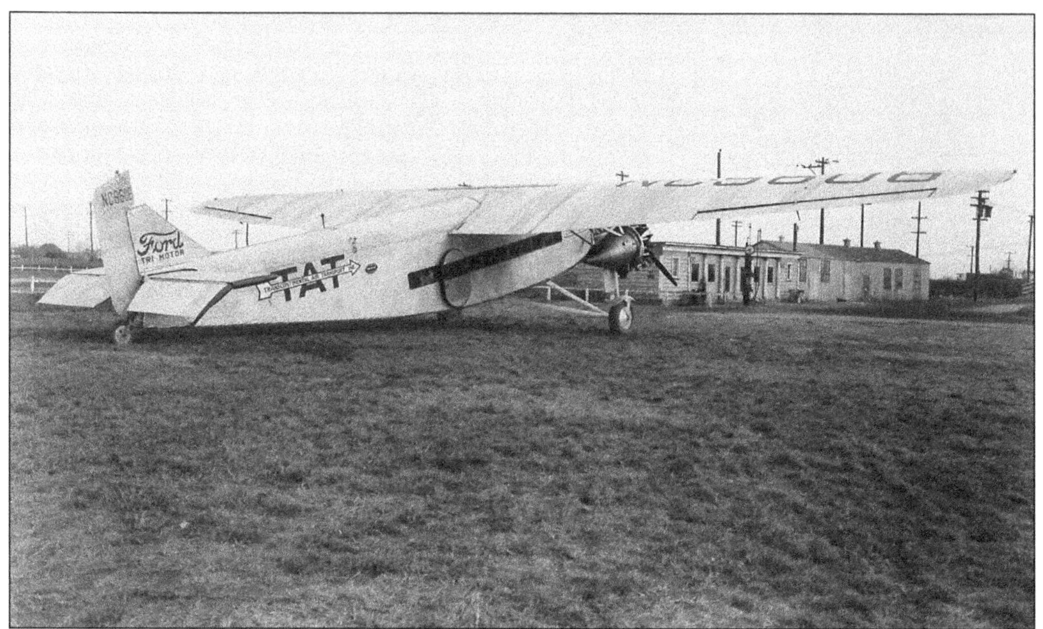

Seen in the 1930s, this Transcontinental Air Transport Ford 5-AT-B airliner, named *City of Columbus*, was fortunate to land at Whander Field. In April 1928, a 10-passenger Pitcarin K-78D plane aborted a landing at the airfield as hundreds of citizens anticipated the first such large aircraft landing, because the crew felt the field was "unsuitable." The field became mostly unused 10 years after its initial opening before being reopened again in 1937. Eventually, Transcontinental Air Transport became Trans World Airlines (TWA). (Courtesy of the Berks History Center.)

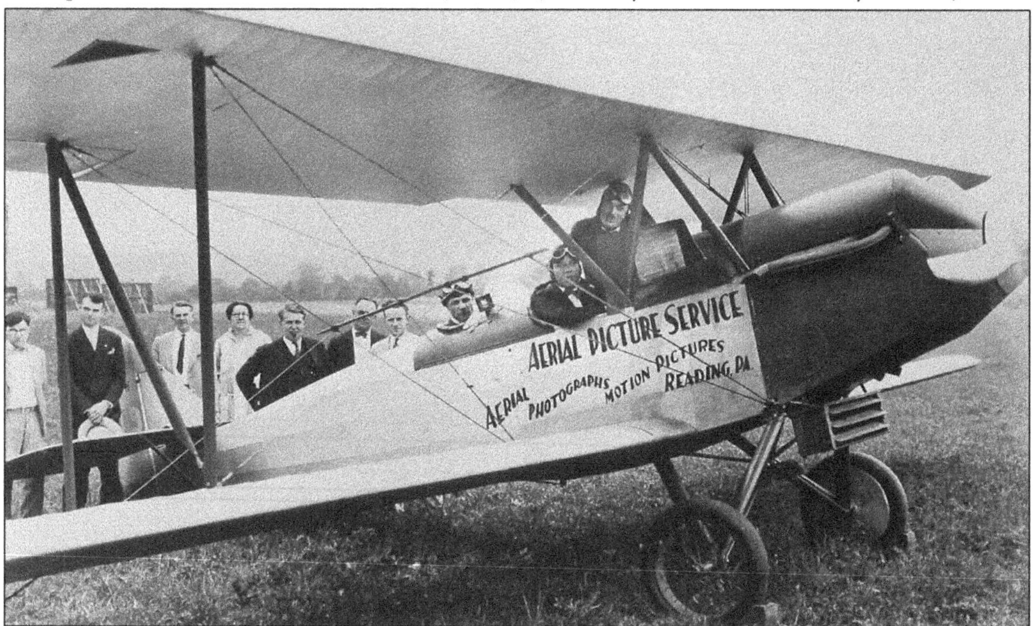

Aerial Picture Service's new Swallow OX-5 is shown at Whander Field in the 1930s. The company performed aerial photography and motion picture services in the area. It was mentioned in an October 2, 1926, *Reading Times* article when it took aerial photographs of the newly constructed Reading Hospital from 2,000 feet. (Courtesy of the Berks History Center.)

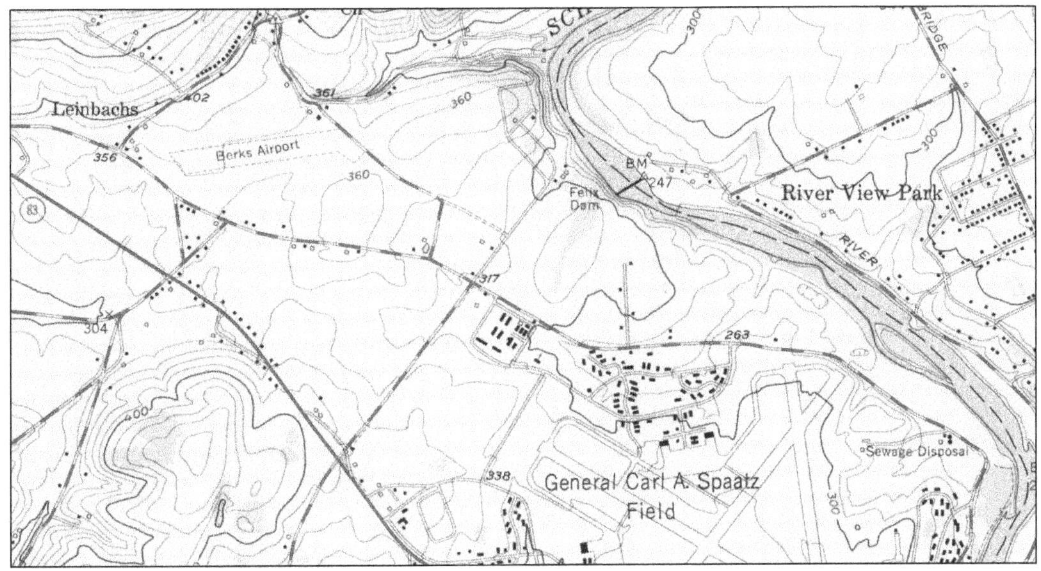

Another farm, approximately one mile north of the present Reading Airport and near Leinbach's Hotel, became the origin of Berks Airport. A farmer, Joel Boyer, had a six-acre flat strip of land that looked suitable for a runway. He leased the land to the Reading Aero Club in September 1937 after the club realized the need for a bigger airport—Madeira Field could not accommodate the growing number and types of planes, with its longest runway being 2,000 feet. Throughout 1937 and early 1938, the club constructed a 2,400-foot runway and seven hangars. (Above, author's collection; below, courtesy of Penn Pilot.)

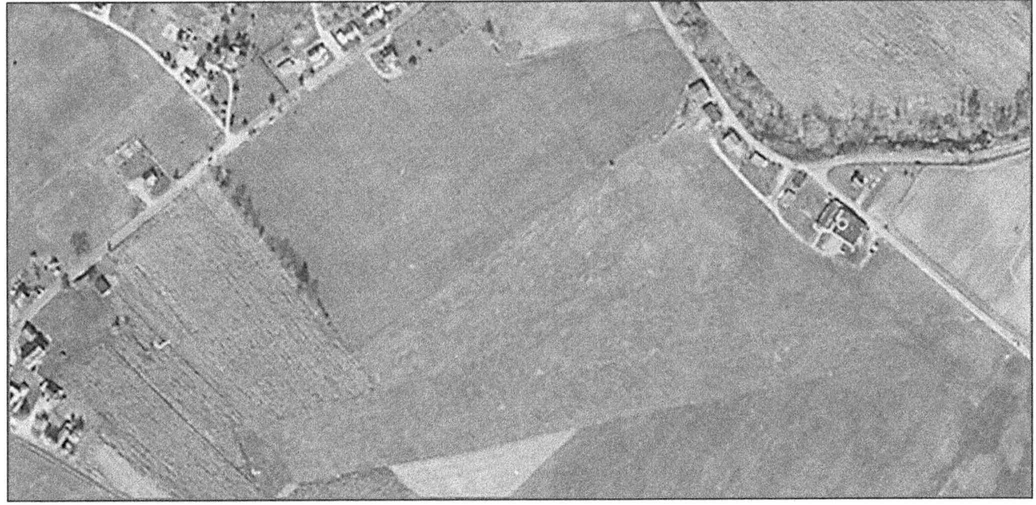

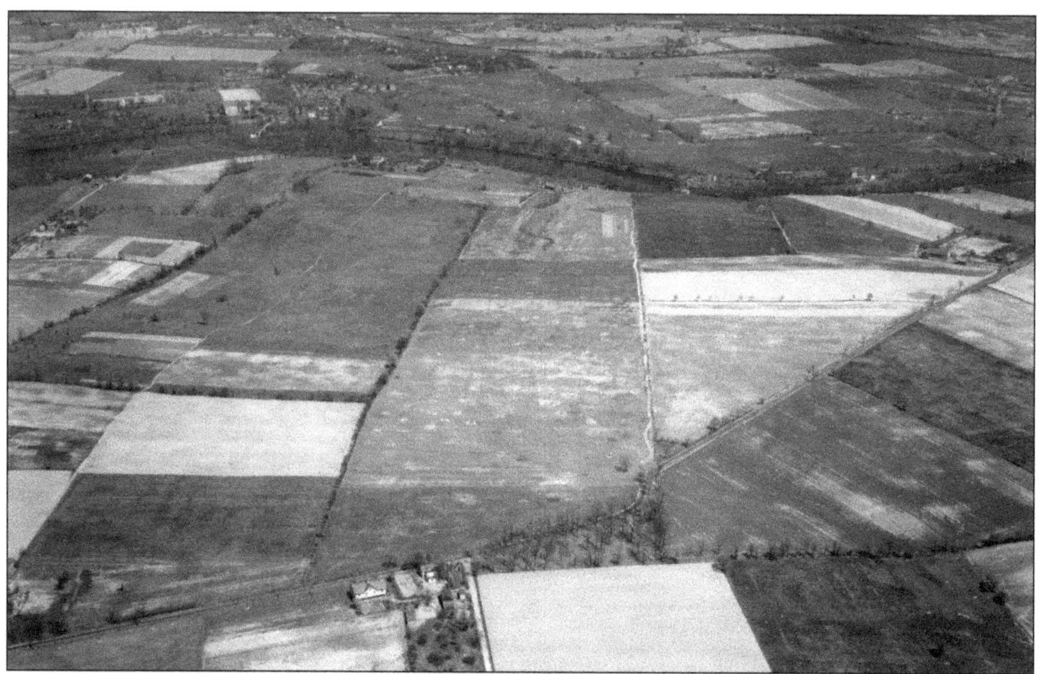

For many years prior to the construction of the Reading Municipal Airport, many pilots and citizens wanted a modern airport, but to develop such a field, estimated at a minimum cost of $500,000, was not possible. It was not until the latter part of 1934 and early 1935 that a committee began investigating options to build the airport, which was around the time that the federal government, thinking of national defense, began allotting money for airport construction through the Works Progress Administration (WPA). It is pictured above prior to construction around 1936, and below around 1944 with its three runways. (Both, courtesy of Terry Sroka.)

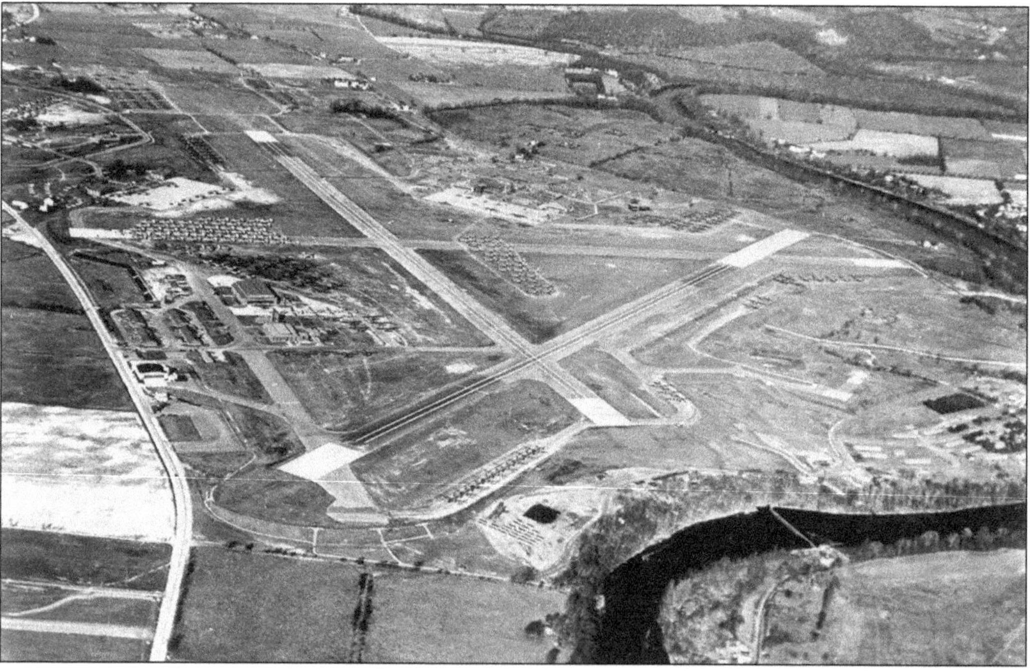

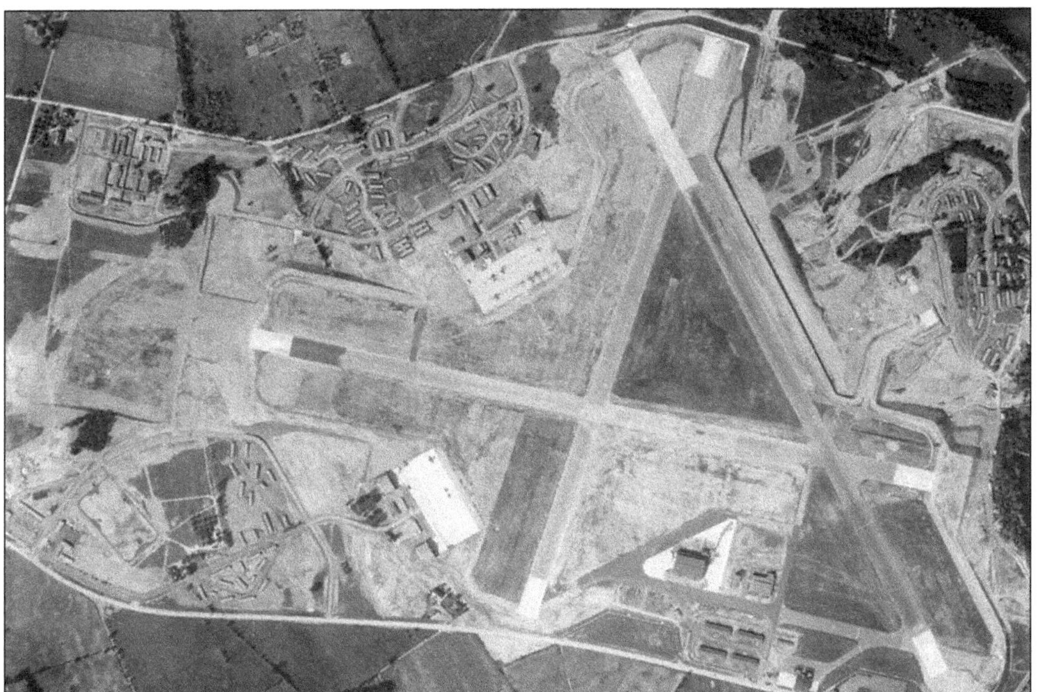

Once a site was chosen in 1934 after considering 18 locations within the county, the city planning engineer was asked to develop the field's layout, runway orientations, runway widths, and parking areas for the proposed municipal airport. On December 10, 1935, county commissioners passed a resolution to pay half of the $15,000 land option costs and other costs for the 180-acre Shearer-Althouse tract. (Courtesy of Terry Sroka.)

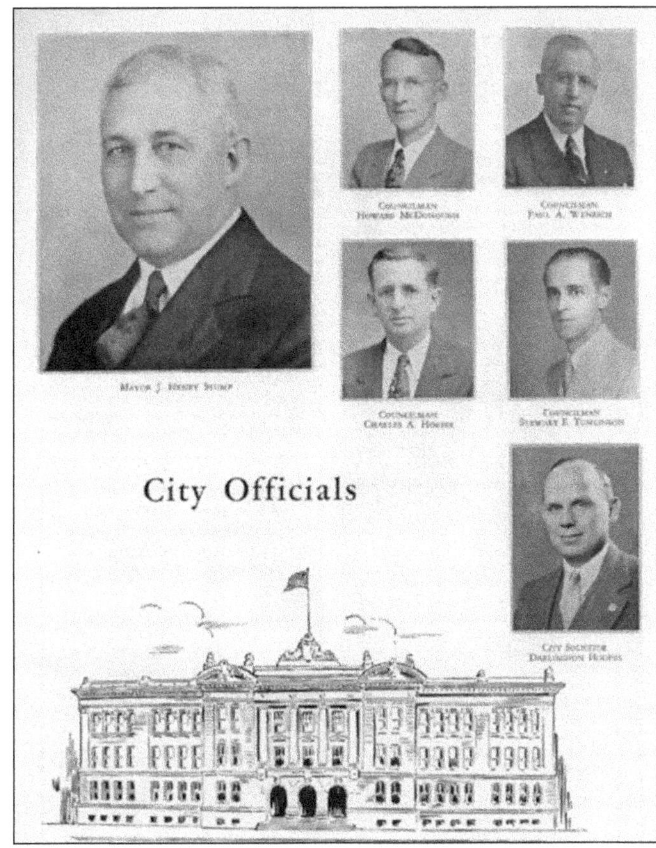

Concern existed that the newly elected city administration would not support construction plans for a new airport. Fortunately, on December 6, 1935, mayor-elect J. Henry Stump issued a statement favoring the airport and Washington approved the application on December 18, 1935, with construction beginning in August 1936. (Author's collection.)

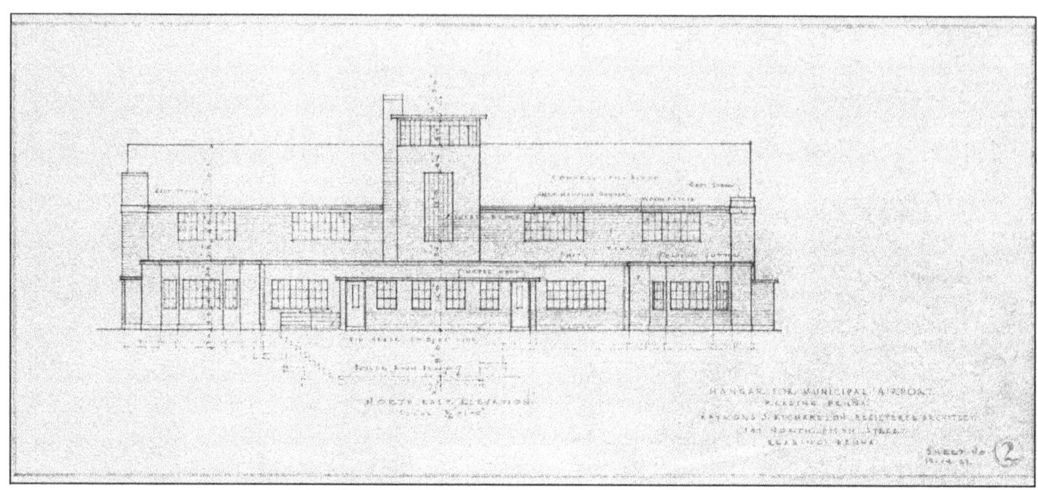

In early 1936, construction delays were experienced due to engineering disputes. The WPA threatened to withdraw the airport project, but on August 6, 1936, ground was broken. During the construction, additional structures were included in the original plan, including a hangar with an attached tower and building at an estimated cost of $44,482. Above is the architect's rendering of the structure, while below is the hangar prior to the dedication. The hangar does not yet have its beacon, which once resided on top of Pomeroy's department store and was gifted to the city. (Both, courtesy of Terry Sroka.)

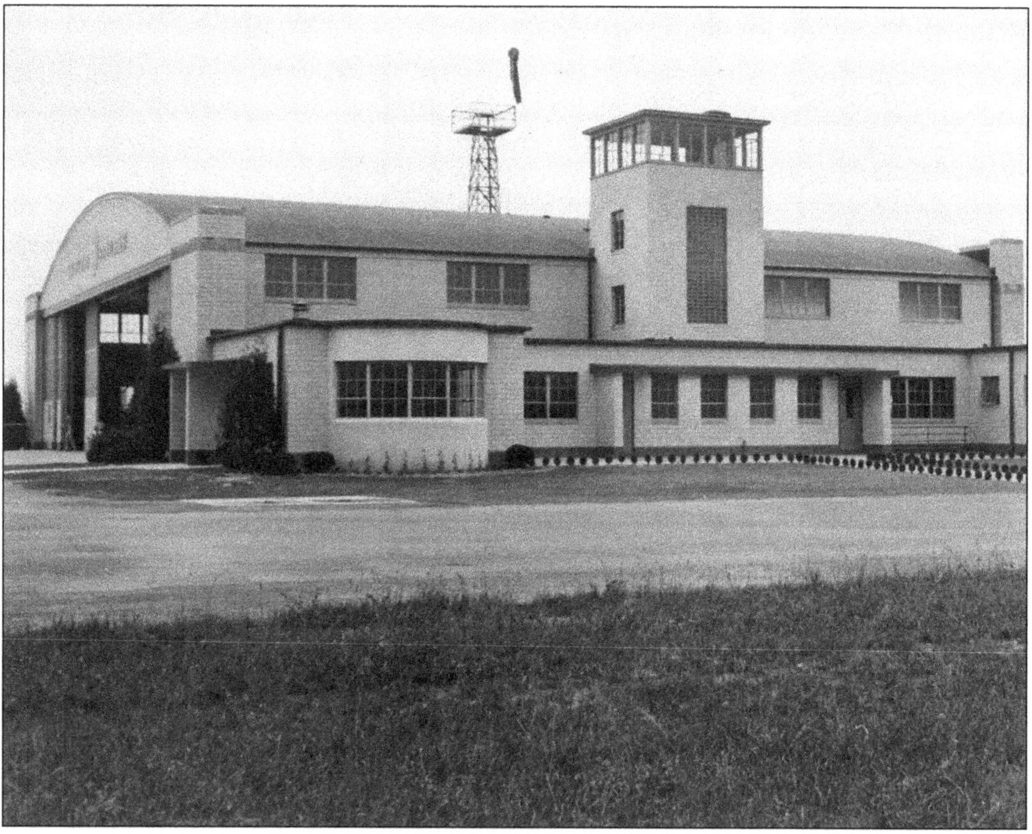

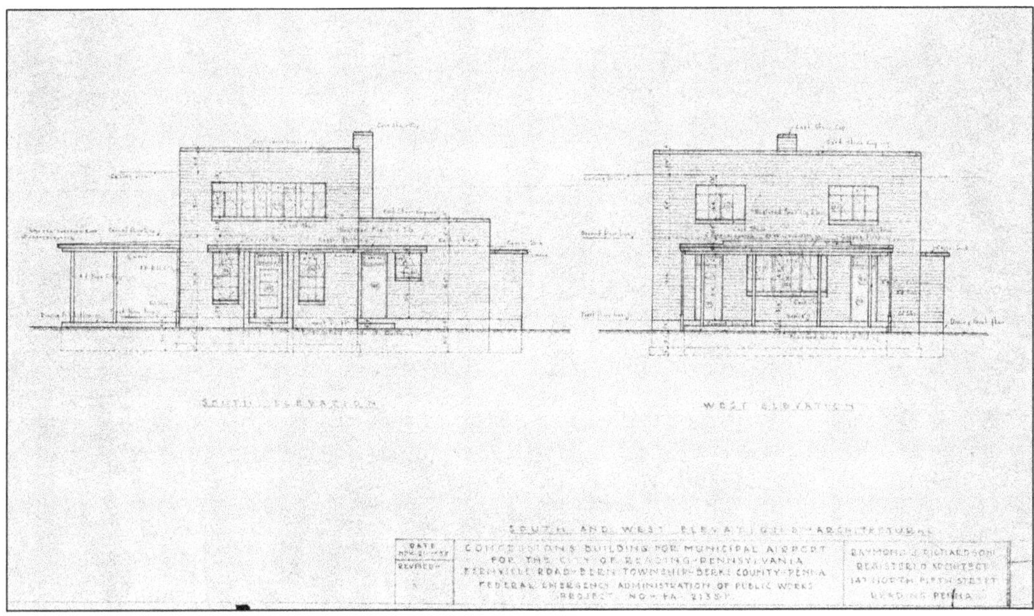

Above is a rendering of the airport restaurant. The photograph below of the restaurant was taken in the late 1940s. The upper level of the building was the residence of the airport operator. The restaurant served food from the semicircular extension. It was a good location to watch aircraft takeoff and land on the field and it still exists at the airport; however, it has been significantly modified from its original design. (Above, courtesy of Terry Sroka; below, courtesy of the Berks History Center.)

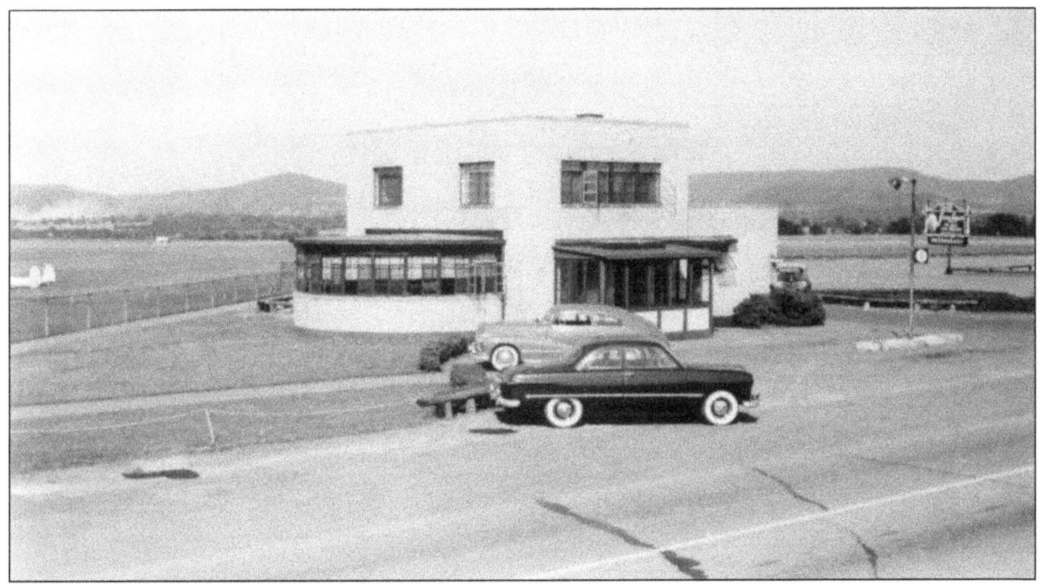

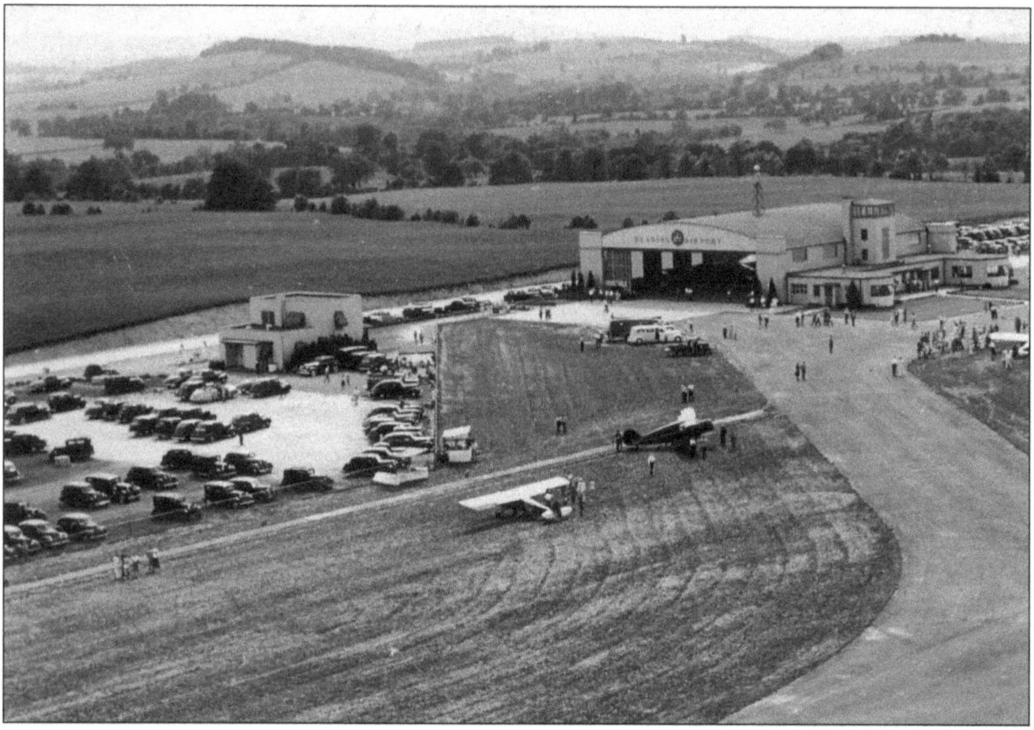

The above aerial photograph was taken during the dedication of the Reading Airport in 1939. Aircraft are on display, and the airport restaurant is shown near the main hangar. During the dedication in October 1939, traffic was so congested that 20,000 cars parked along highways, fields, hilltops, and dirt roads within a six or seven-mile radius just to see the air shows. Buses attempted to relieve some traffic. (Both, courtesy of Terry Sroka.)

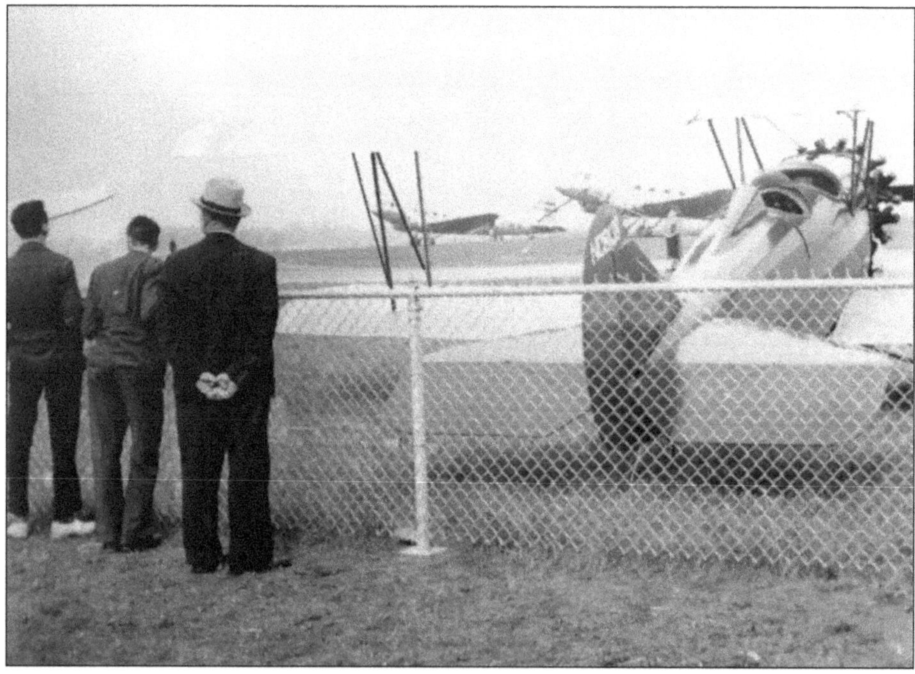

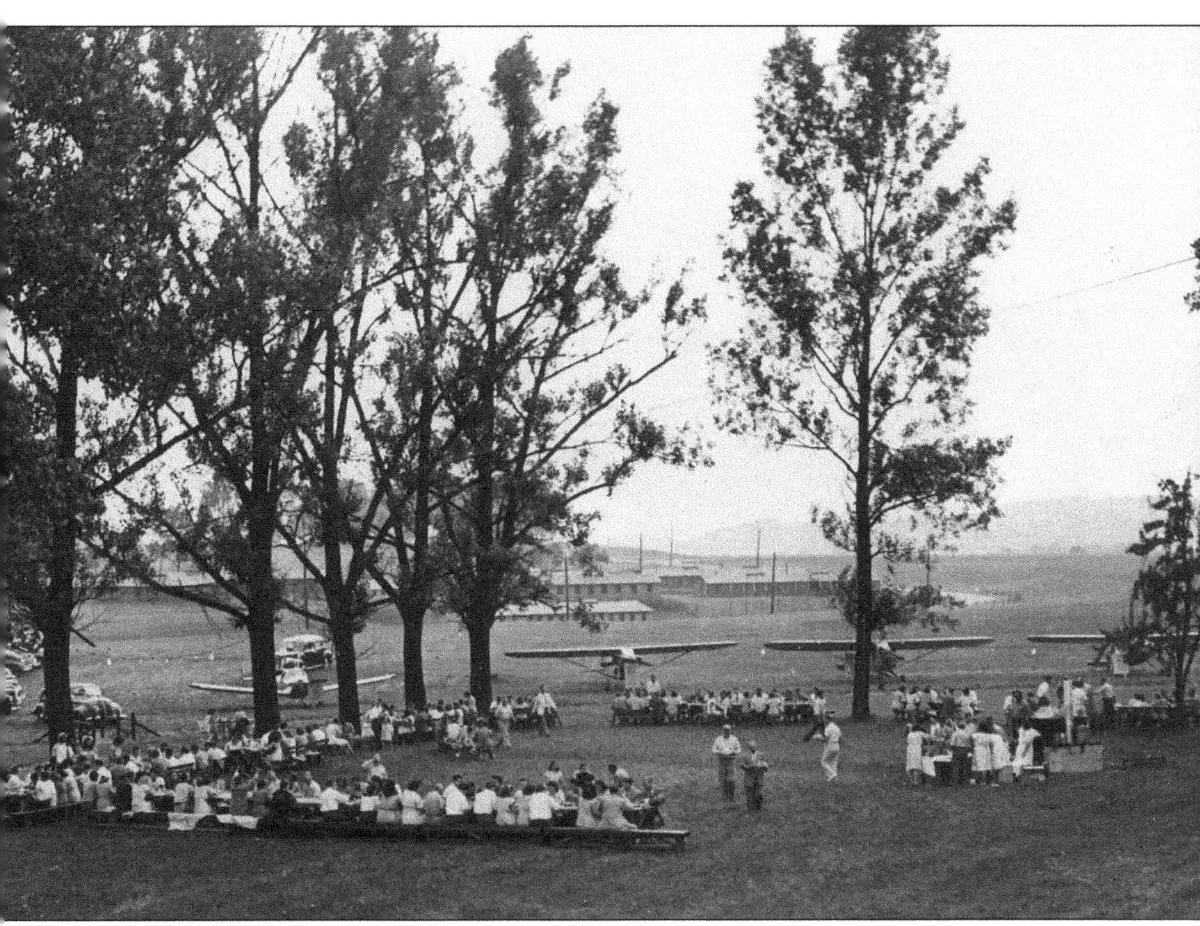

Special events for family and friends of airport personnel and the military were held at the airport throughout the years. This scene is on the west side of the airport during the late 1940s, with aircraft lining the perimeter of the picnic grove. In the background, barracks that originally housed German POWs were eventually converted into affordable residential housing units after World War II had ended and the Army no longer had a use for them. (Courtesy of Terry Sroka.)

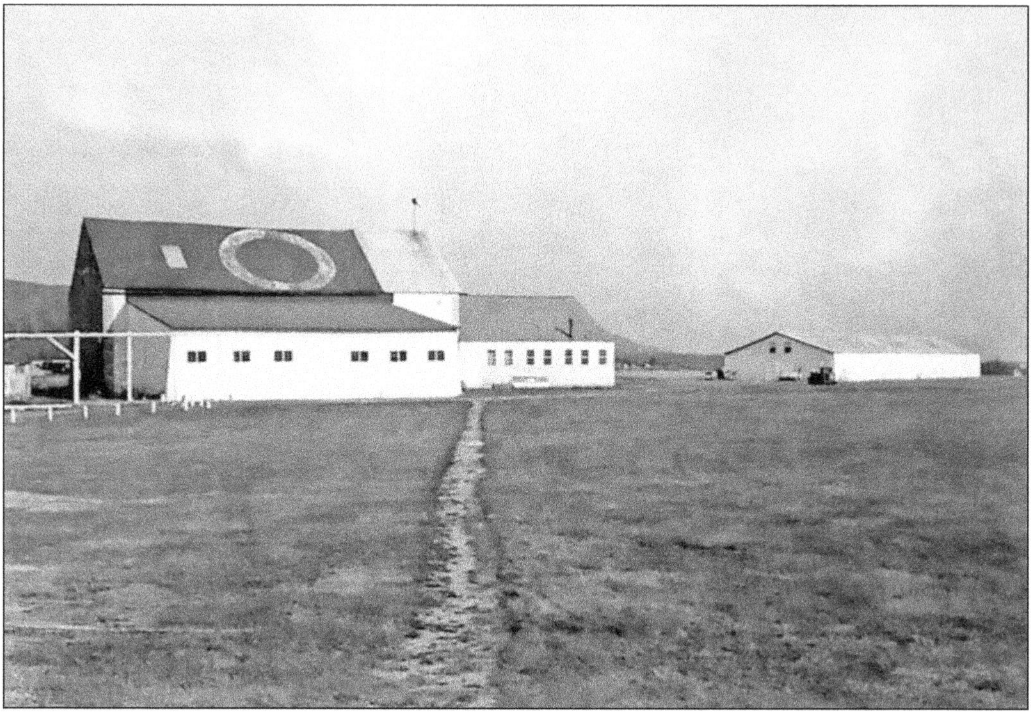

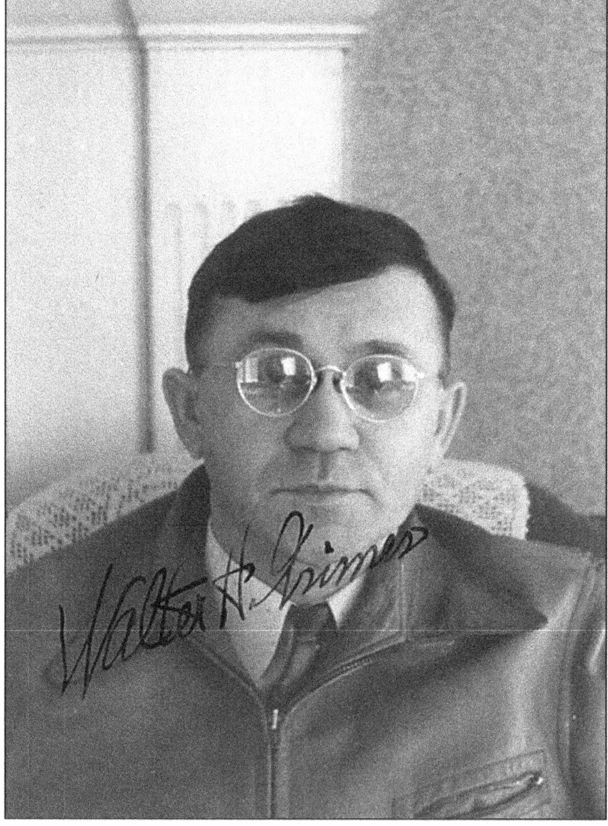

Prior to creating Grimes Airport, Walter H. Grimes purchased 119 acres in 1937 and operated a grass airfield at the northeast intersection of School and Marion Roads in Marion Township, north of Stouchsburg, Pennsylvania. Three hangars and a shed were built on the airfield to accommodate approximately six aircraft, mostly Piper Cubs, Taylorcraft, and Fleet biplanes. The airfield was in existence until June 1945, when it was sold when Walter Grimes began operations of Grimes Airport in Bethel, Pennsylvania. The photograph above shows the airfield marking "Open Circle," designating Grimes Airport. Originally, the airport had two 3,000-foot runways, but it now has only one 2,720-foot turf runway oriented in a slightly southeast/northwest direction as a result of extensive Interstate 78 construction through the property. (Both, courtesy of John Grimes.)

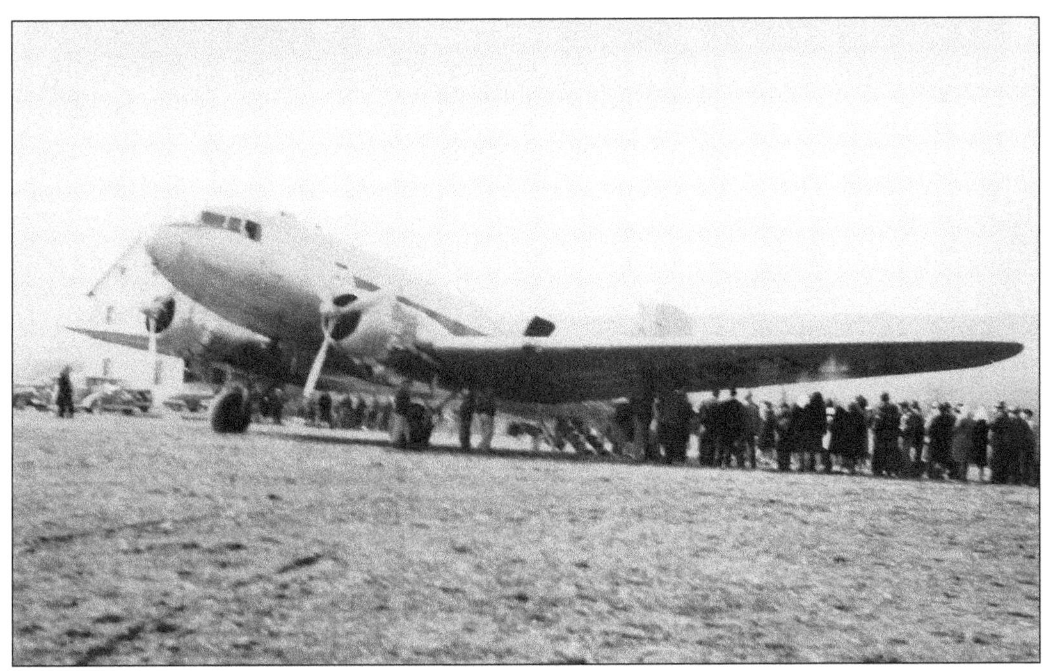

Years ago, Grimes Airport was a location for flight instruction, along with charter and maintenance operations, with turf runways large enough to accommodate Douglas DC-3 aircraft, as shown above. The airport has been fortunate not to experience the fate of many other early fields. This was the result of a vision and commitment to preserving earlier aviation through the current Golden Age Air Museum's operations at the airfield and the ongoing aviation-related events still held for the public. Below is a Piper Cub near the airport's smaller hangar, with two men observing its operation. (Both, courtesy of John Grimes.)

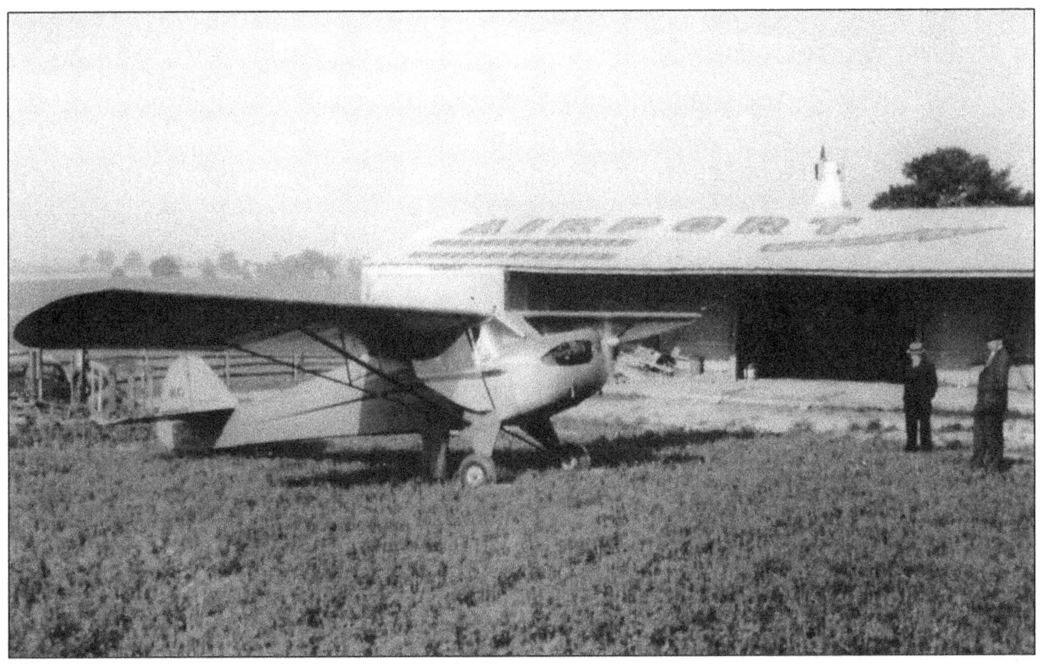

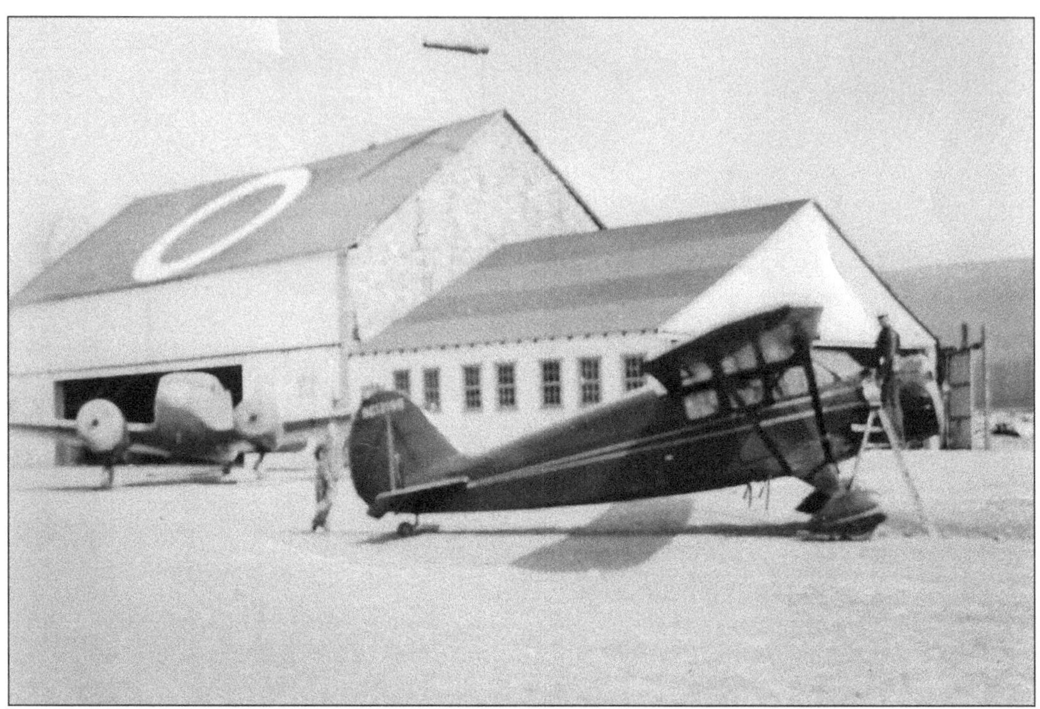

On the field at the main hangar above is a Cessna T-50 "Bamboo Bomber" in the background while someone prepares Walter Grime's Stinson, used for commercial purposes. Below is a 1959 Piper Commanche, with David Grimes standing proudly in front of the aircraft. The name of the Cessna T-50 was a result of it being made of wood and tubular steel covered with fabric. It was a popular twin for the civilian and commercial light transport market. (Both, courtesy of John Grimes.)

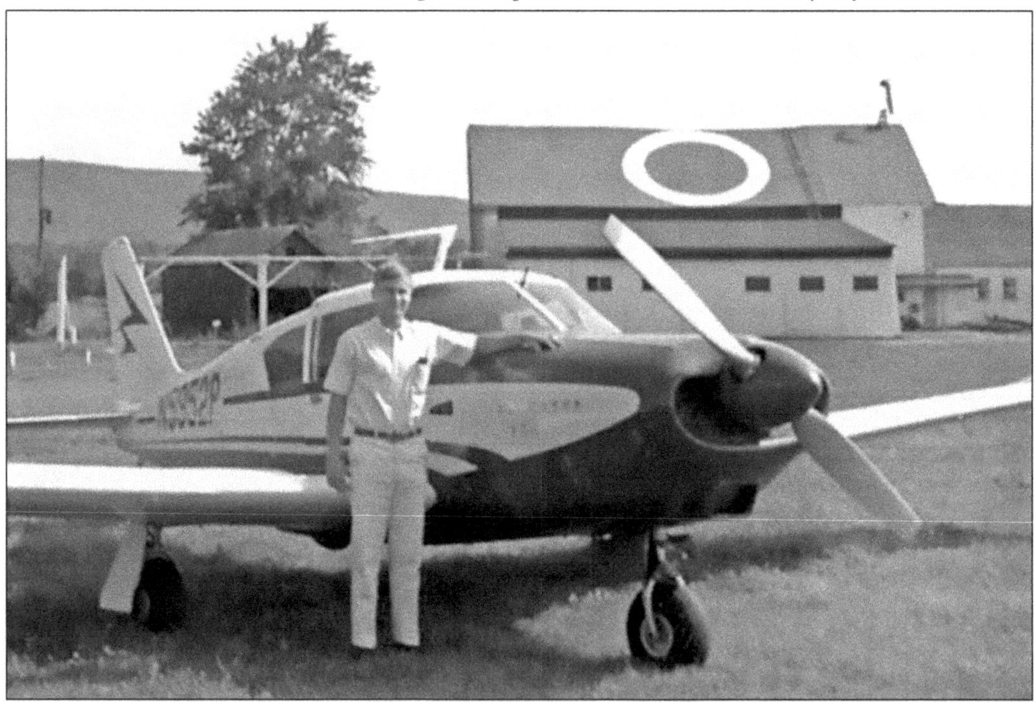

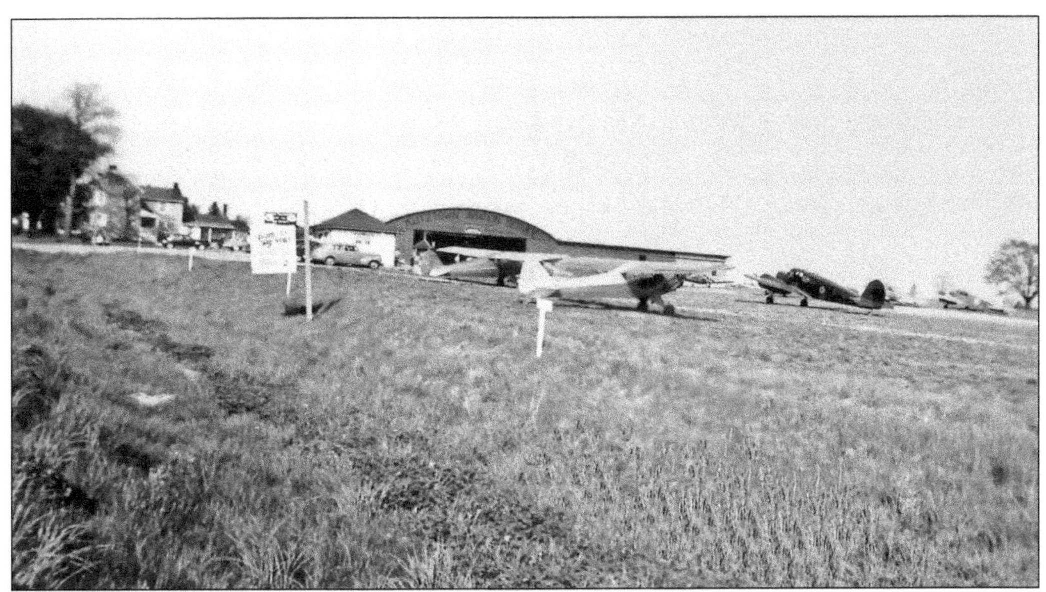

Alton, Clarence, and Warren Stein created Kutztown Airport in the 1920s, but it officially opened in 1945 with Fritz Bachman as manager and instructor. The airfield, pictured above in the 1940s, closed in 2009. It faced numerous potential closures as early as 1971 before owners of the Kutztown Airport Diner purchased the property, and in 1973 when local businessmen kept it in operation. The airport had two runways—an asphalt runway and a turf runway. Trailer homes were on the property, as shown below in the 1970s. Kutztown University purchased some of the property for university parking. (Above, courtesy of Tom Yenser; below, courtesy of Friends of Kutztown Airport.)

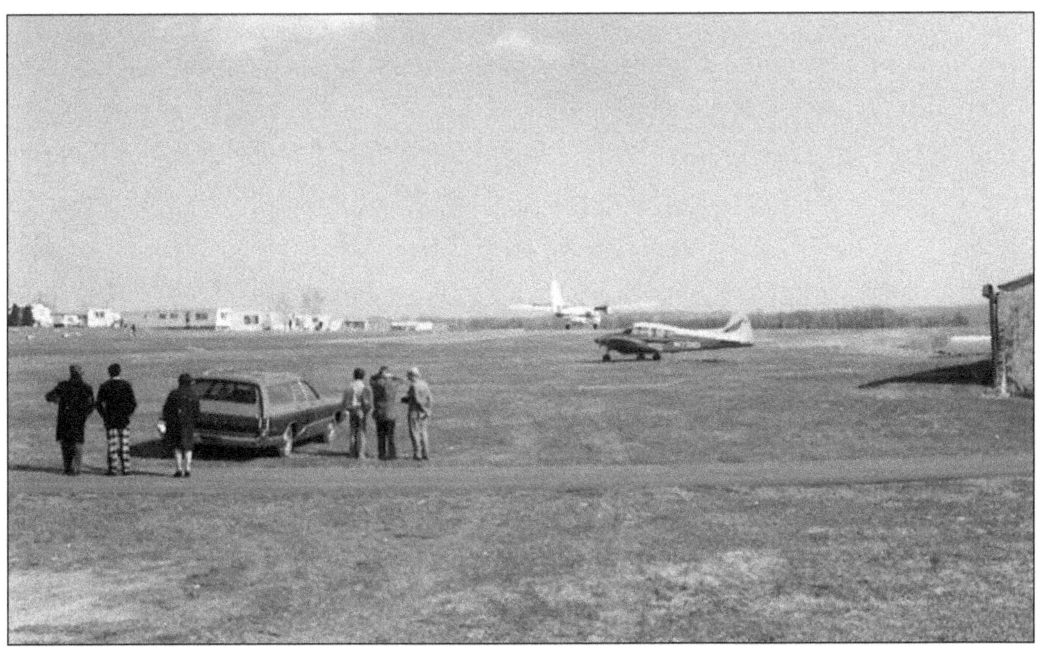

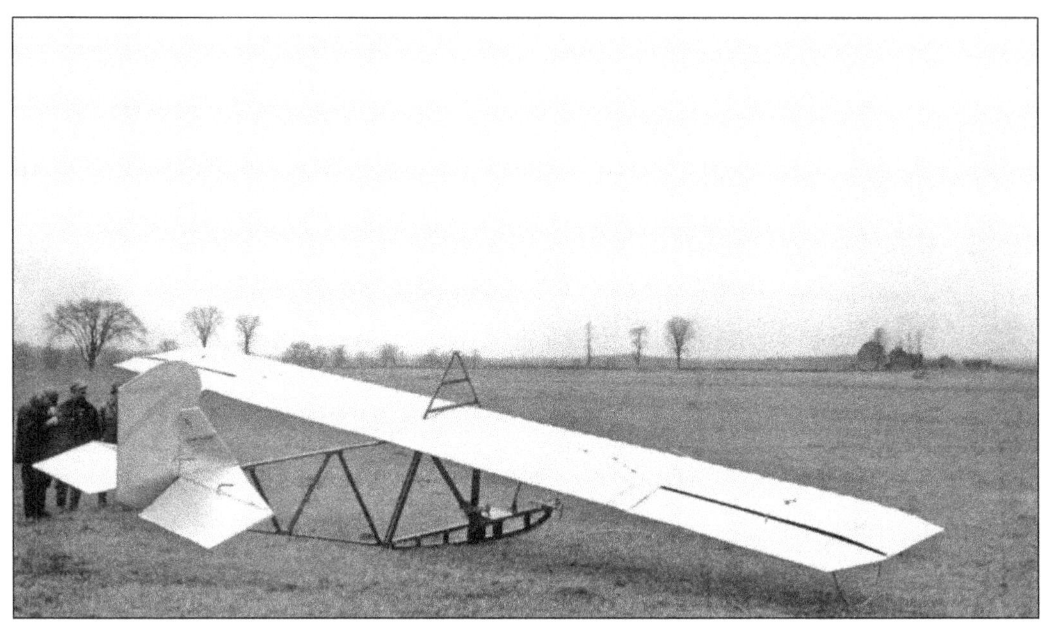

Gliders were an important part of Kutztown Airport's Pennsylvania Dutch heritage. Numerous articles were written that mention glider contents at the airport with the Soaring Dutchmen Club's Fersommling—a Pennsylvania Dutch social event in which food was served, speeches were made, and plays or skits were performed for entertainment. These photographs are of Clarence Stein's *Mead Primary Glider*, with him at the controls. (Both, courtesy of Tom Yenser.)

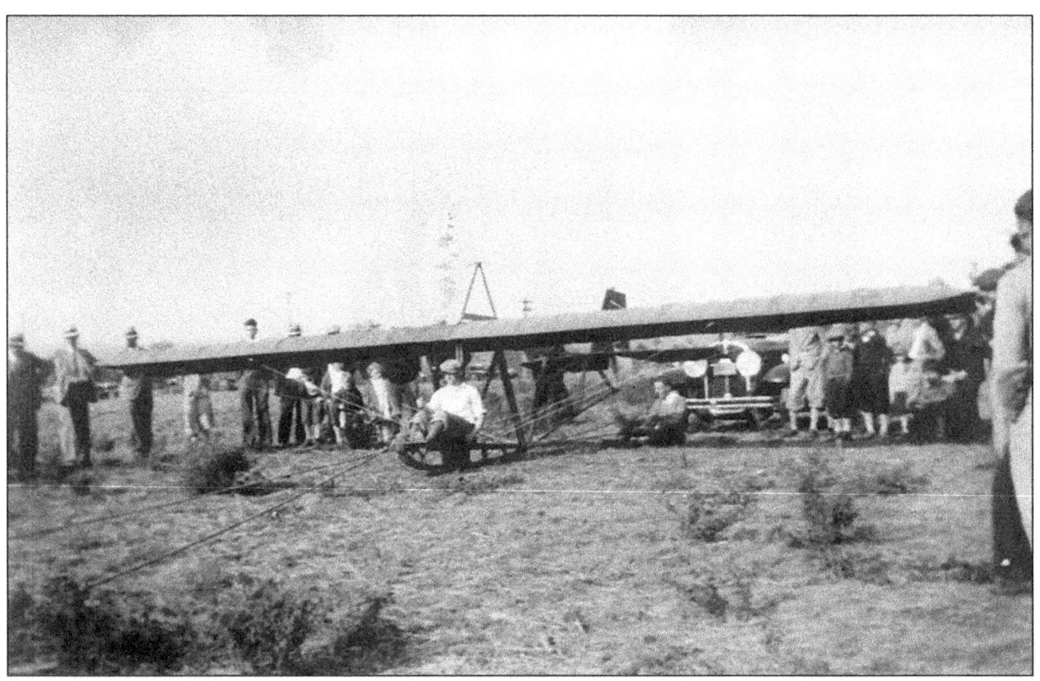

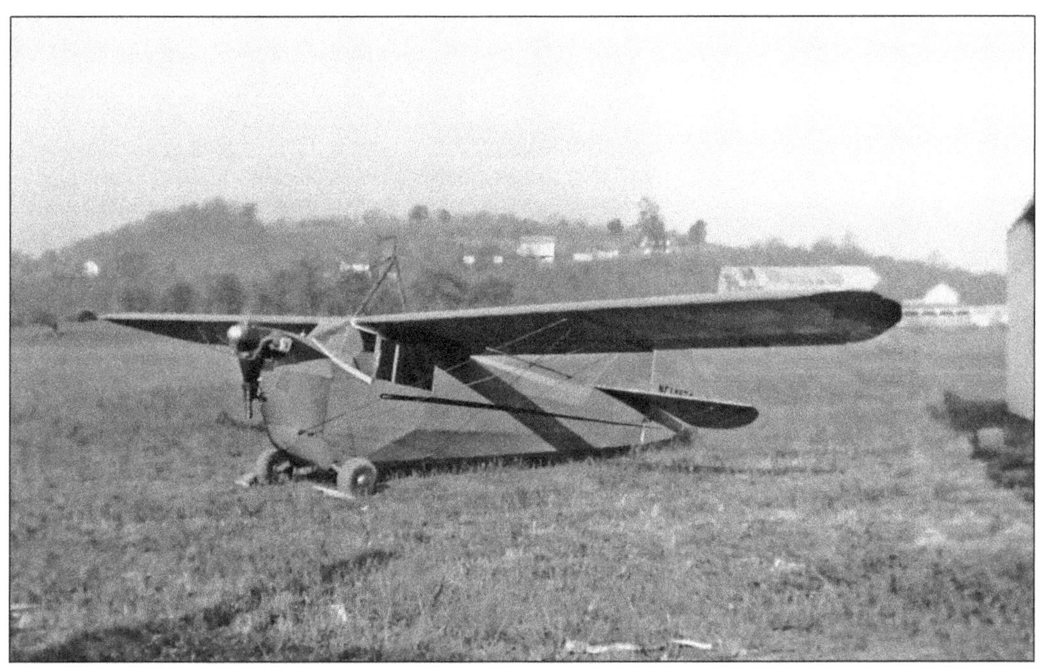

Powered aircraft were also popular at Kutztown Airport, as seen in the above photograph taken in the 1930s of an Aeronca C3 and the below picture of some young boys riding an old Buick while Clarence Stein looks on in front of his aircraft. The Aeronca C3 was dubbed the "Flying Bathtub" because of its unusual fuselage shape. Its 40-horsepower engine made it an extremely slow but economical airplane to operate. (Both, courtesy of Tom Yenser.)

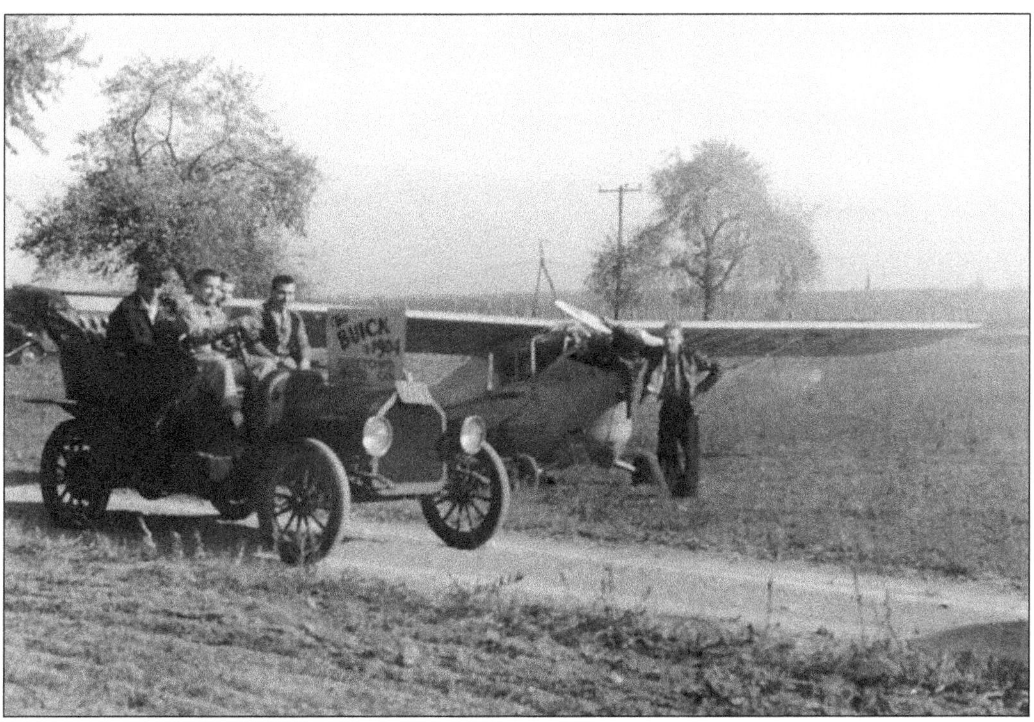

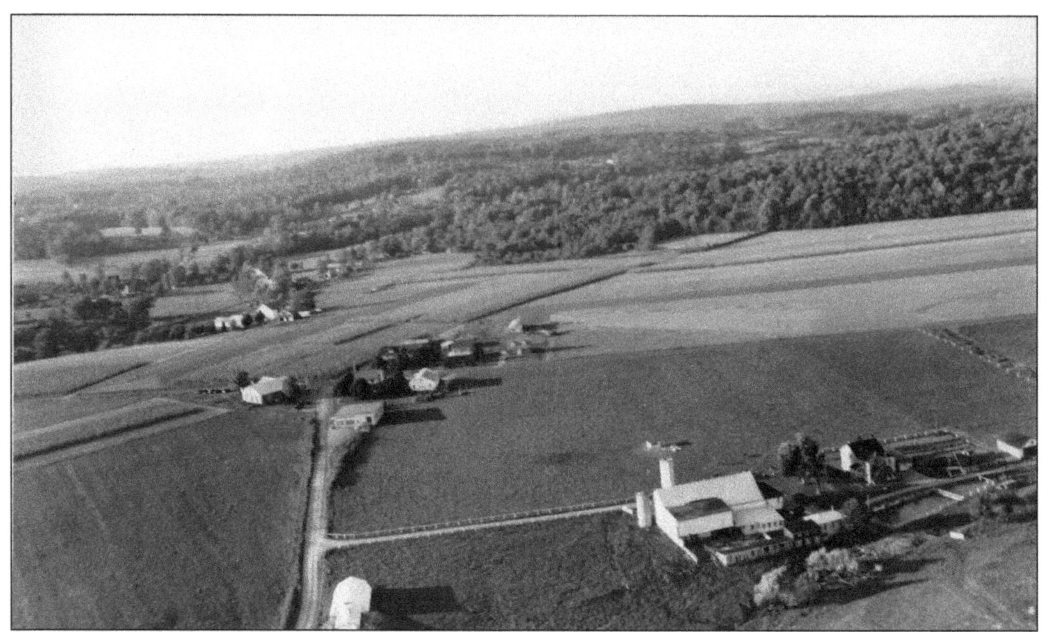

Morgantown Airport, a privately owned and public use airport, is a 2,600-foot grass field, which officially opened in 1947. The picture above was taken prior to the 1950 construction of the Pennsylvania Turnpike, with the airstrip in the background and Levi Beiler's farm in the foreground. Today, Black Diamond Soaring provides most aerial activity at the airport. In 2002, the field faced closure when Pennsylvania Power and Light wanted to purchase the land. Instead, over 50 residents and pilots opposed the purchase, and the owner decided the airport benefitted the community. In 1955, a veteran operator of Reading Airport, Melvin Nuss, considered the airport site as a Reading and Berks County jet airport and in the late 1960s, plans were evaluated to build another airport in the area to accommodate the growth of local businesses and industry. (Both, courtesy of Paul Beiler.)

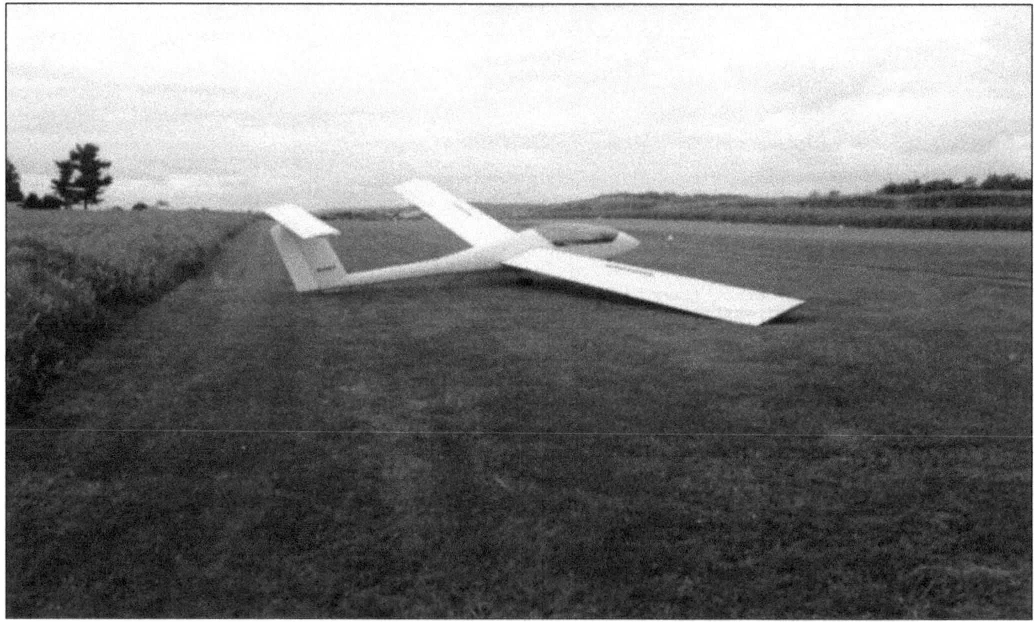

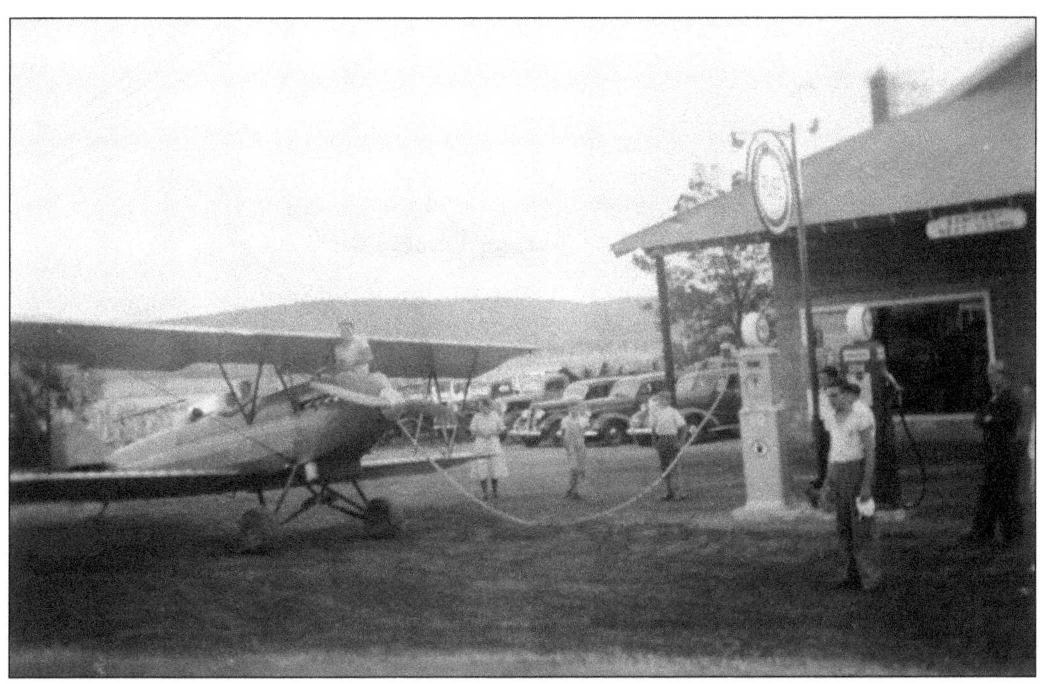

These photographs were taken at Beiler Motors in 1938–1939 in Morgantown, Pennsylvania, where David Beiler had a gas station and car dealership close to the turf airstrip. A Pietenpol Air Camper has pulled up for fuel; the aircraft's Model T engine ran on auto fuel. (Both, courtesy of Ken Stoltzfus.)

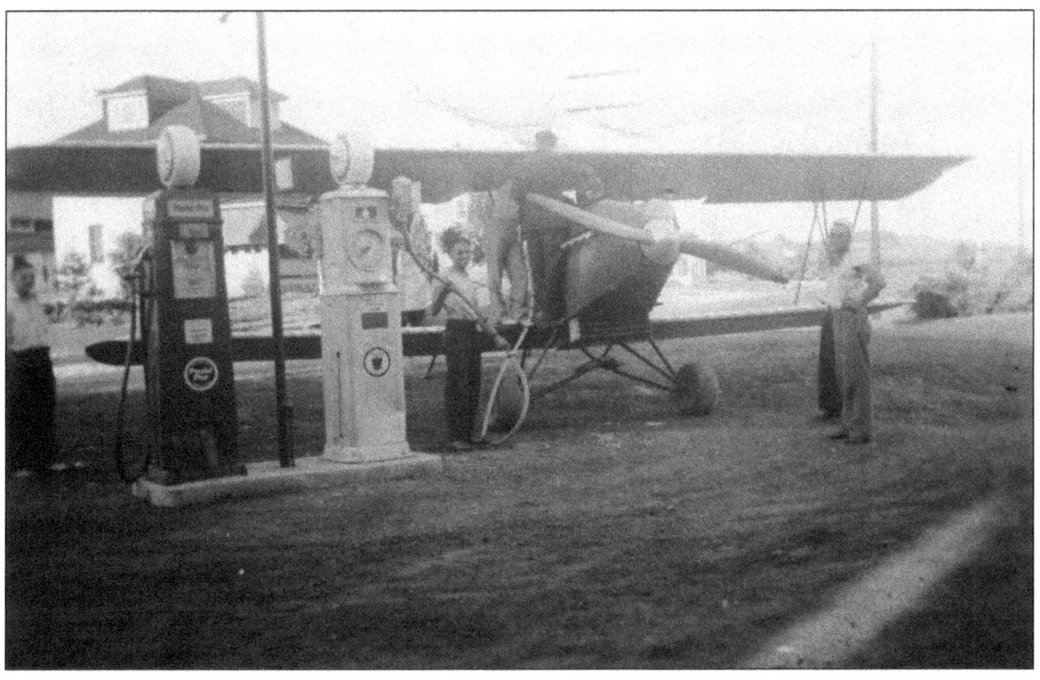

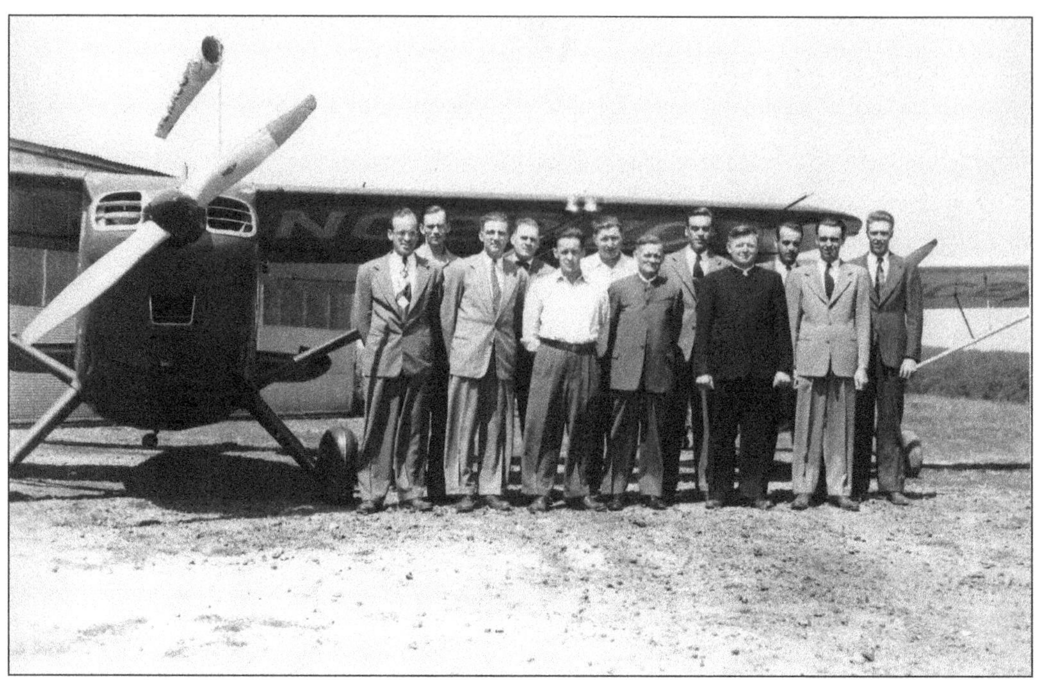

Members of the Morgantown Aero Club are pictured around 1946 when the club was founded as a nonprofit corporation. Above, the Stinson Voyager, along with an Aeronca Champ in the background, was used for many purposes, with an emphasis on helping local church workers. These photographs were taken in front of a hangar on the farm of Levi Beiler. Pictured from left to right are Leroy Stoltzfus, Mahlon Stoltzfus, David Hertzler, Ralph Hertzler, Simon Zook, Paul Yoder, Levi Beiler, Clarence Stoltzfus, David Beiler, Ralph Stoltzfus, Emory Stoltzfus, and Raymond Beiler. (Both, courtesy of Paul Beiler.)

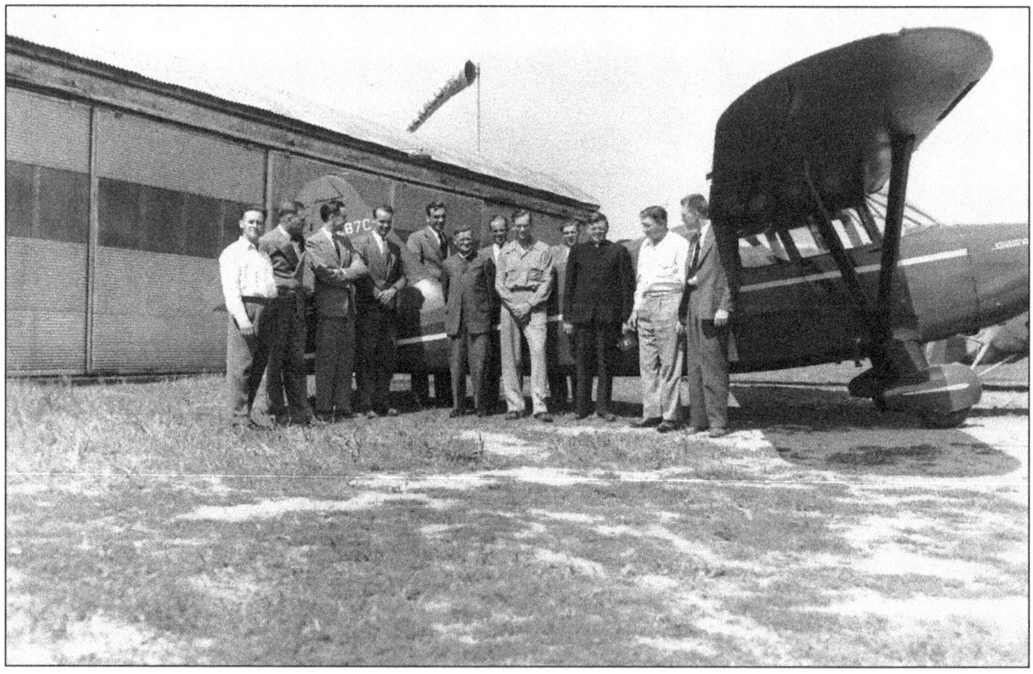

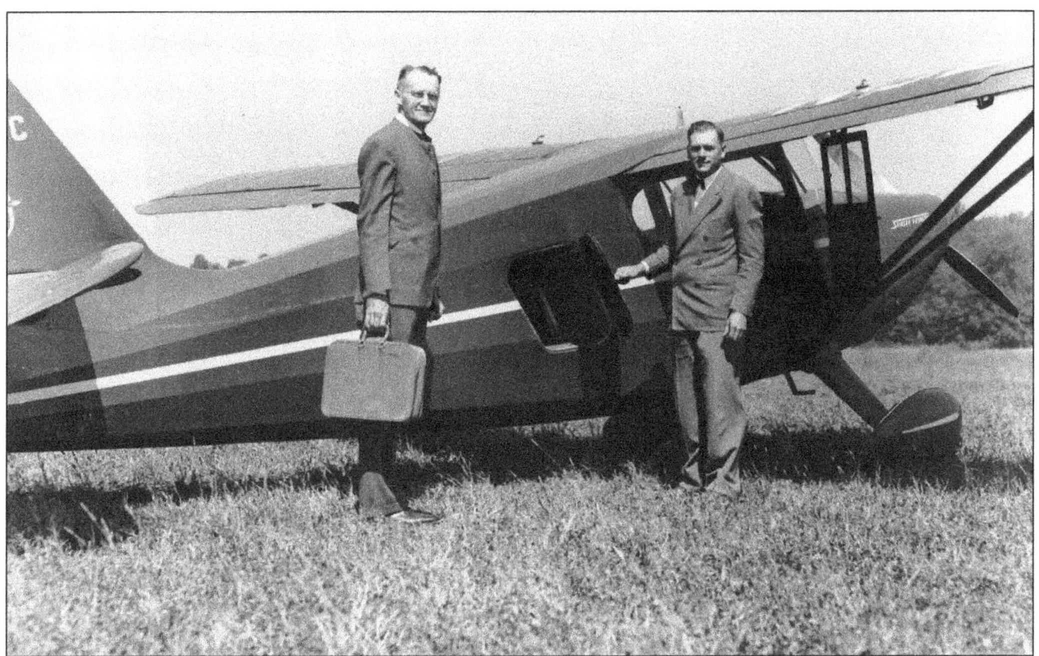

William Detweiler (left), a minister from Orrville, Ohio, is ready to be flown on a trip to Alaska in July 1948 as Ralph Hertzler stands at the open baggage compartment of the Stinson Voyager. The trip was sponsored by the Mennonite Church to explore the possibilities of missionary work. (Courtesy of Paul Beiler.)

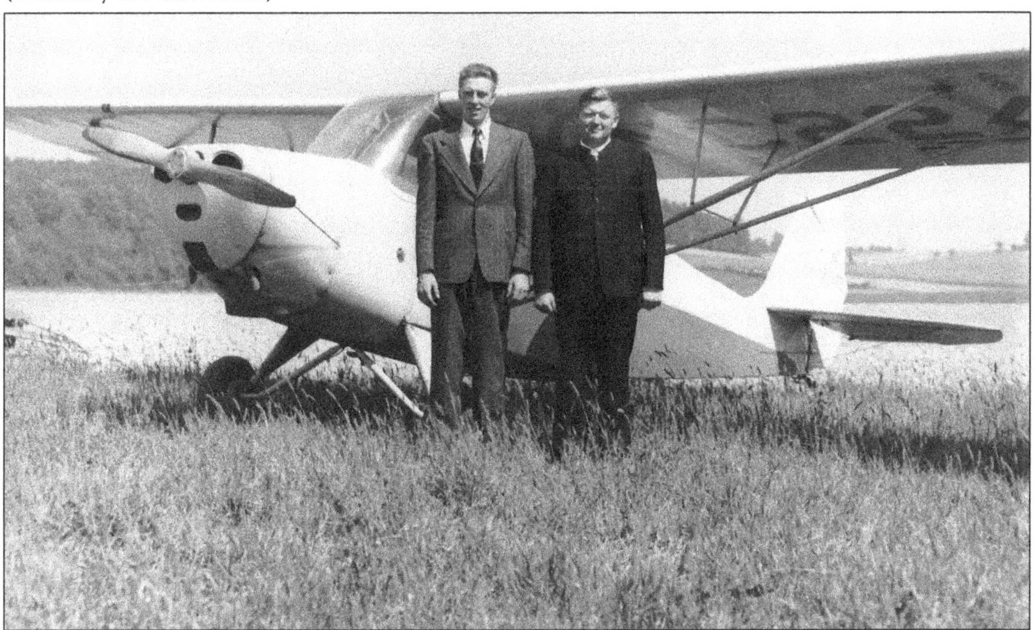

Raymond Beiler (left) and David Beiler stand in front of an Aeronca Champ at Morgantown Airport in the late 1940s. As members of the Morgantown Aero Club, they owned a cabin in the mountains of northern Pennsylvania and occasionally used the airplane to fly there for fishing and hunting. Ministers were flown to appointments 50 to 100 times per week to assist the church in a variety of outreach events. (Courtesy of Paul Beiler.)

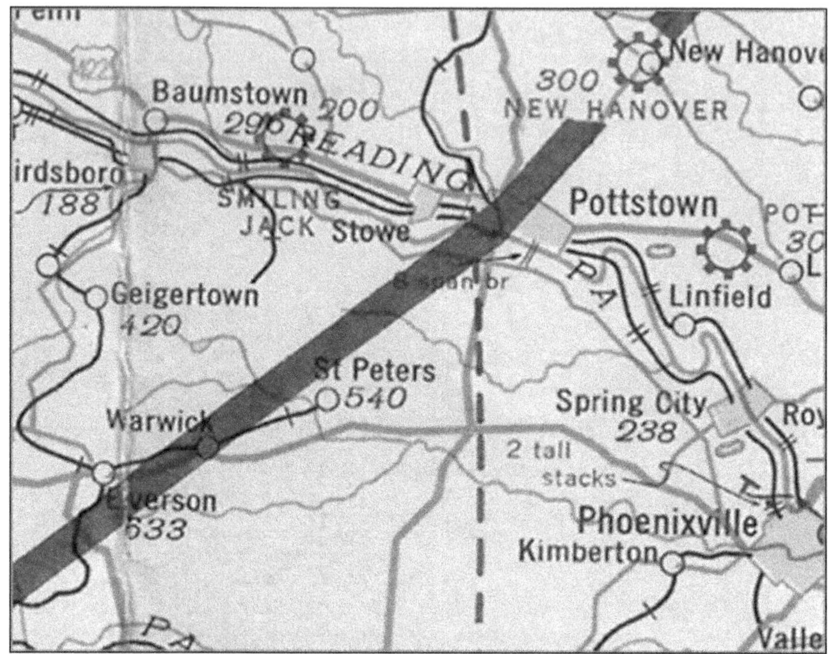

Officially opened on September 5, 1942, Smilin' Jack Airport, named after a popular 1930s comic strip pilot, was located in Douglassville. Initially, it started operating with 11 airplanes piloted by New Jersey and Philadelphia-area residents as a way to continue flying outside of restricted airspace during the war. The airfield, operated by Frank M. Fuhr of Essington, was situated along the north side of a section of Route 422 and occupied 60 acres of a triangular-shaped property having an all-way sod landing area measuring 2,400 feet northeast to southwest. Living accommodations were available adjacent to the airport for those using the airfield. (Above, author's collection; below, courtesy of Penn Pilot.)

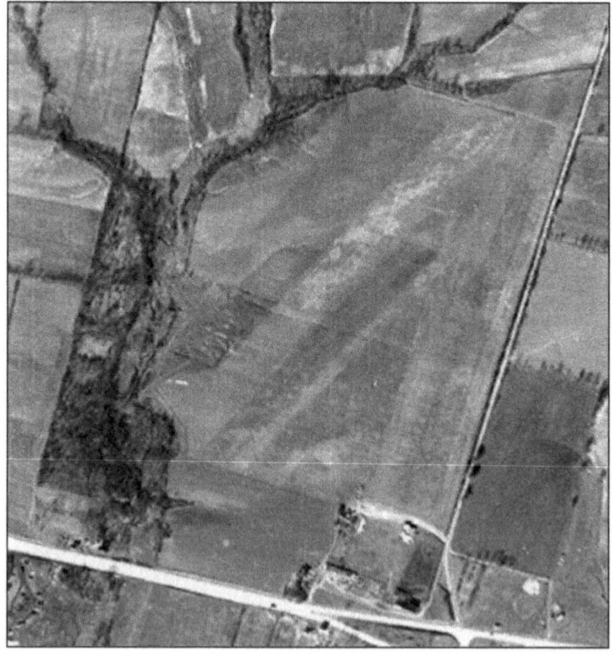

The air marker was the simplest and least expensive aid to aerial navigation, and was the most effective and necessary means for private and non-commercial aviators. The painted aid on the old Reading fairground grandstand indicated airports with orientations relative to a northerly course. Those airports were Reading Municipal Airport (three miles northwest) and Madeira Field (one mile southeast). The markings were simple in design and visible from a height of 3,000 feet. They followed certain formats and dimensions set by the US Civil Aeronautics Administration, as shown in the image below from volume six of *The Passing Scene*. (Right, courtesy of George M. Meiser IX; below, author's collection.)

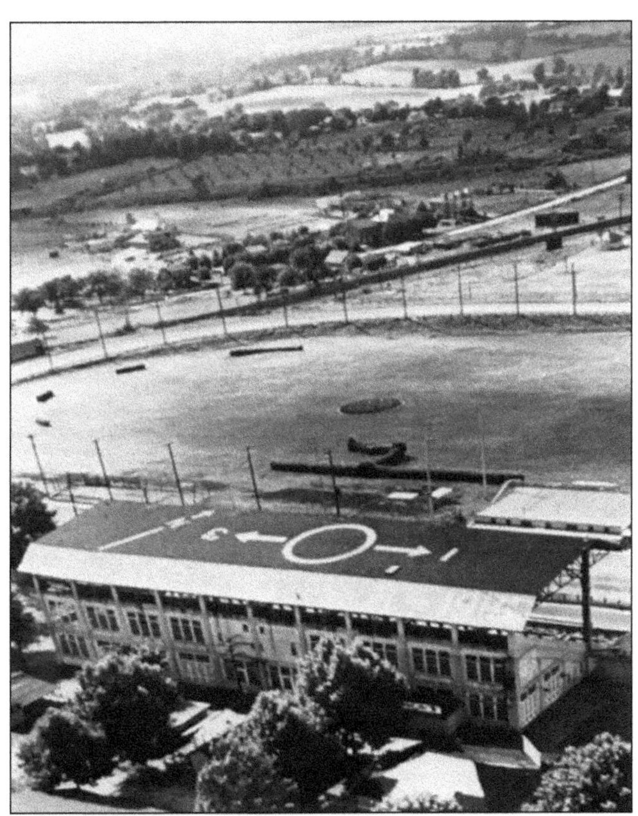

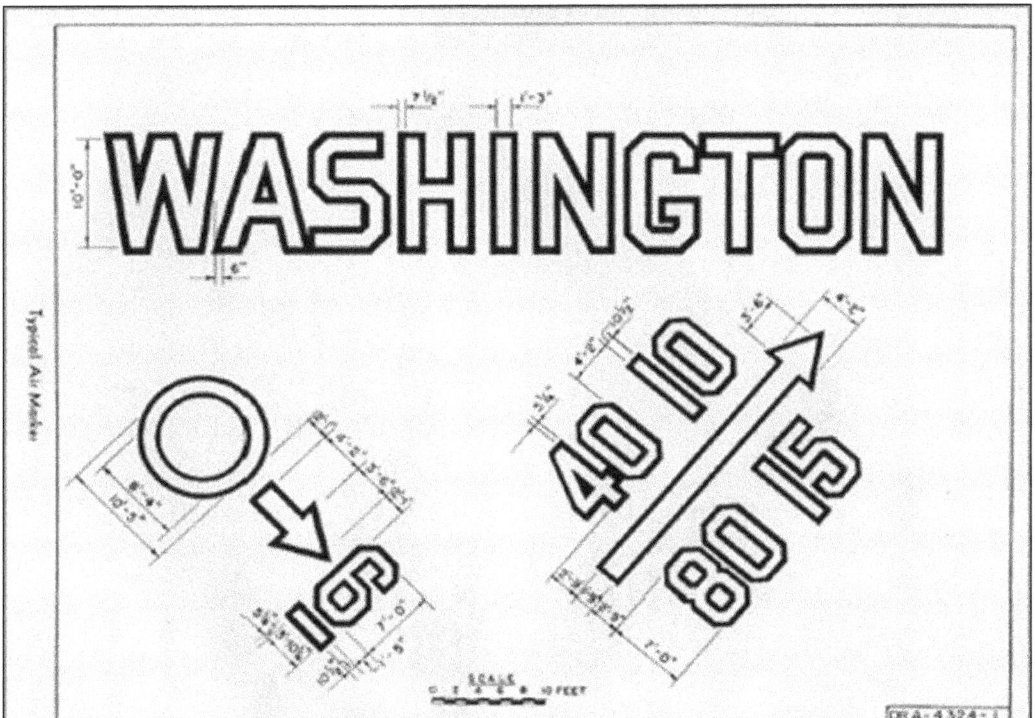

WASHINGTON

Typical Air Marker

40 10

80 15

16

SCALE

10 FEET

DRA-4324-1

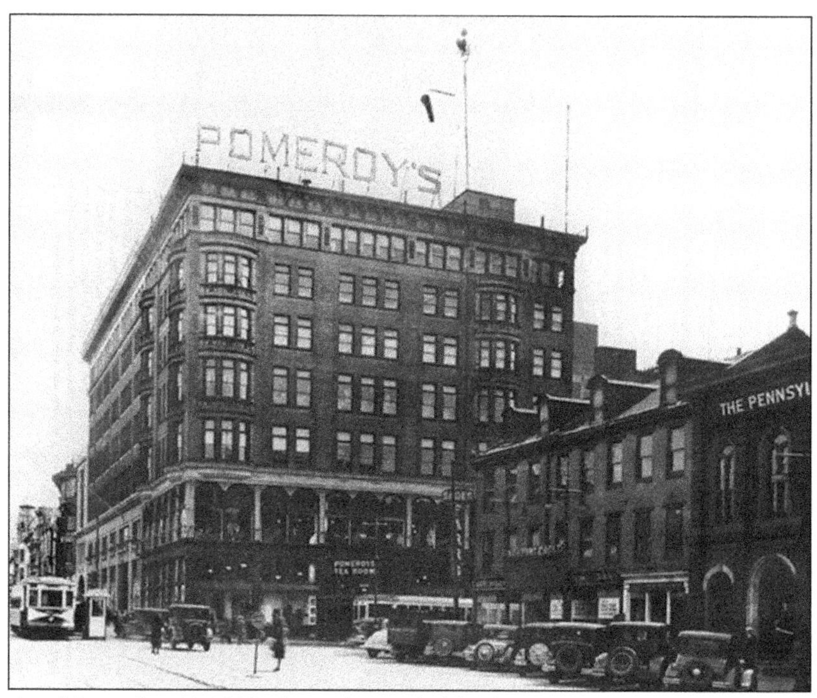

Operational on November 11, 1929, above the roof of the former Penn Street Pomeroy's department store were a windsock and a beacon—a two-million-candlepower light capable of being seen for 40 miles used to guide pilots along a route from Pittsburgh to New York. When the new Berks County Courthouse was built in 1931, the beam shone into the 18th floor of the courthouse; adjustments were necessary to ensure pilots were able to navigate to Whander Field. The beacon is still functional and is now located at Reading Municipal Airport. As seen below, Pomeroy's also had roof markings identifying the direction of the nearest airfield (Madeira, seven miles to the north) and the town (Reading). (Both, courtesy of the Hagley Museum and Library.)

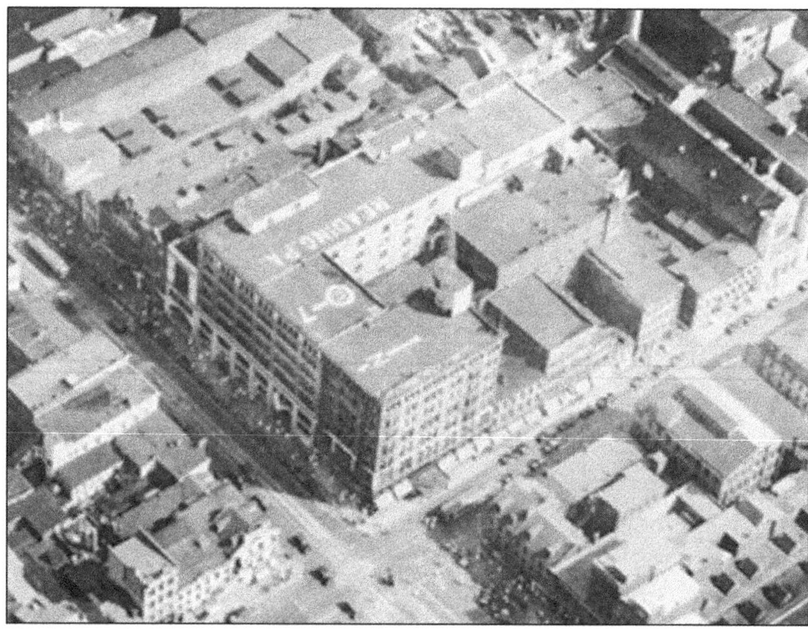

Five

Air Shows and Events

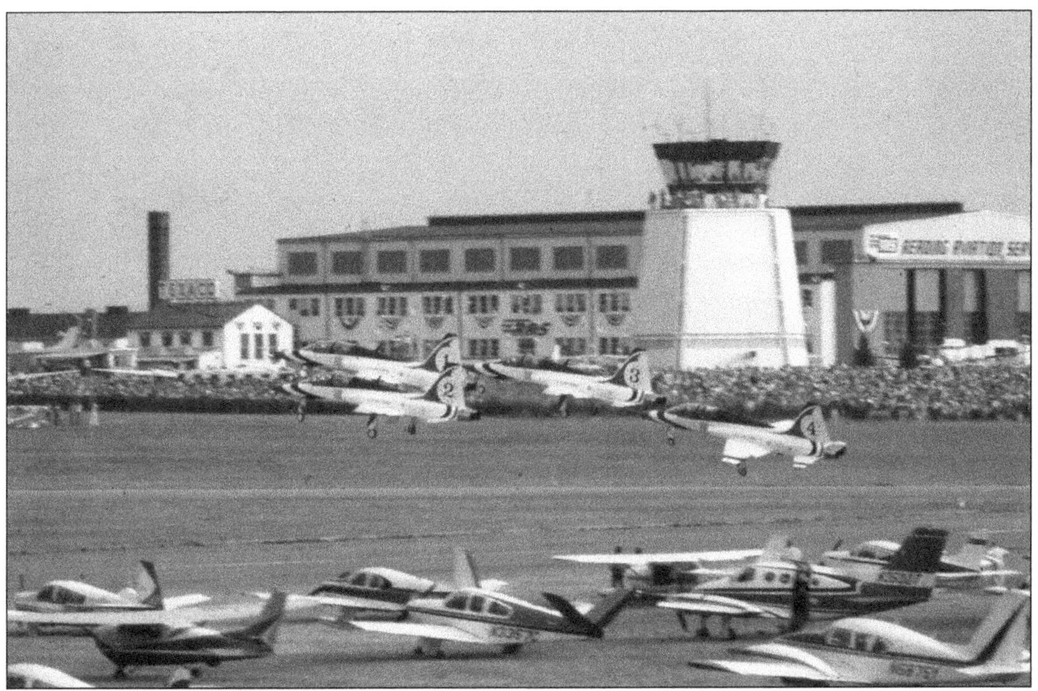

The first air shows at Reading were managed by the Reading Airport Authority in 1939 with the opening of the Reading Airport. After World War II, community flying clubs and the airport held modest events until 1949, when Reading Aviation Service began inviting vendors and customers to Reading. During that time, there were no air shows at what was called the National Maintenance and Operations Meetings, but eventually, scheduled aerial acts became a reality and the event was renamed the Reading Airshow. (Courtesy of Terry Sroka.)

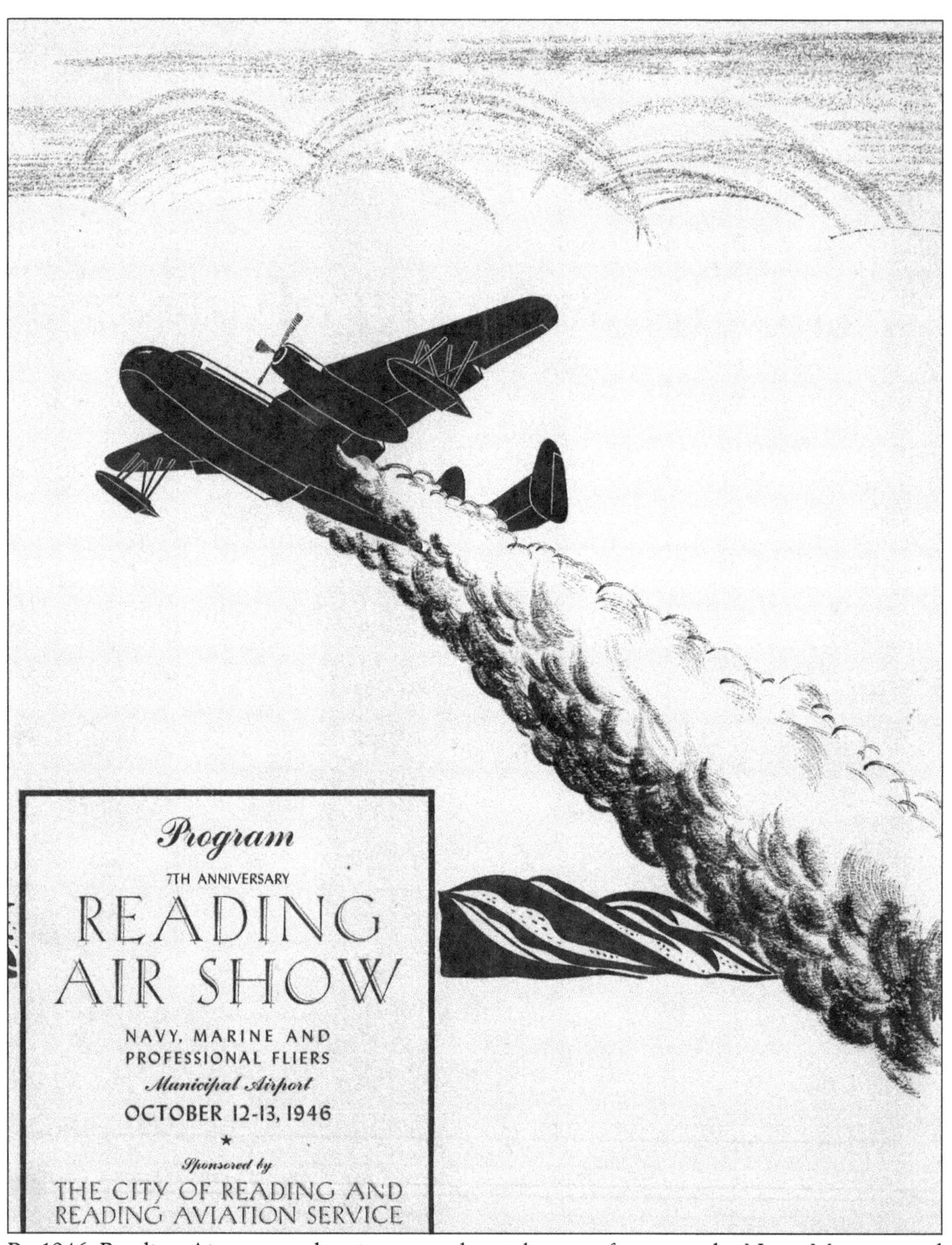

By 1946, Reading Airport was hosting annual aerial events featuring the Navy, Marines, and professional aviators, and had become one of the best air shows in the world. Reading's air show was considered the second largest aviation event in the world. Its initial format was primarily an aviation vendor trade show. Although the aerial entertainment was incidental to the trade show, eventually, it opened and closed the trade show and drew crowds not only from the industry, but also from the local community. At its peak, it had 12,000 exhibitors and brought much business to the local economy. (Courtesy of Terry Sroka.)

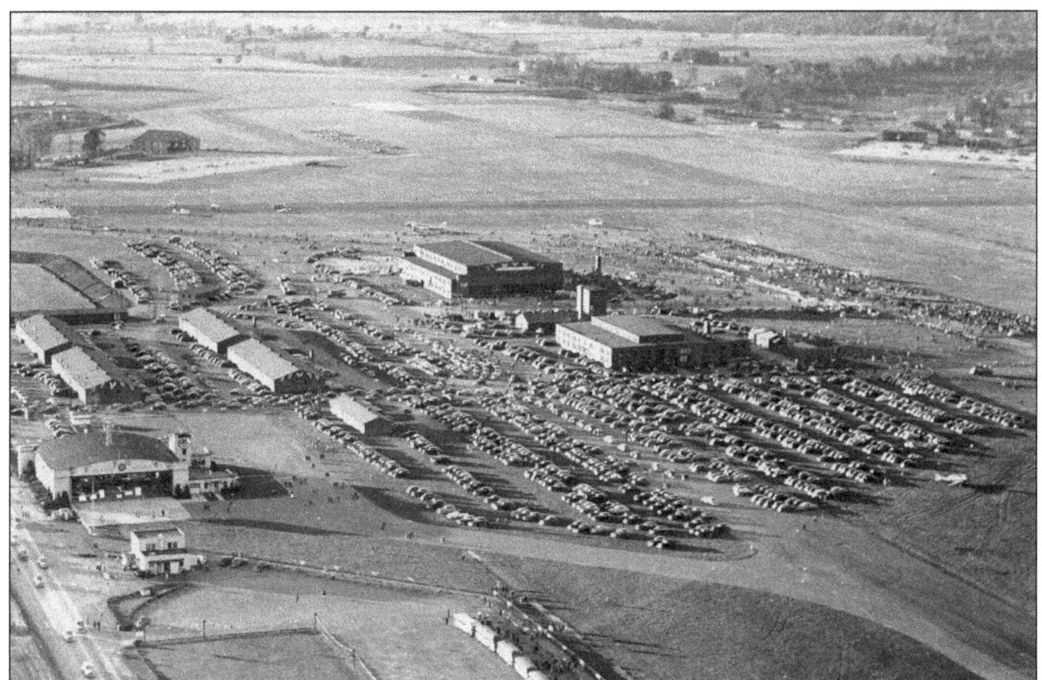

The picture above shows how the airport appeared in the early 1940s when the National Maintenance and Operations Meetings were annual events. By the 1970s, as shown below, the airport and extent of aircraft and structures grew, highlighting how popular the Reading Airshow had become and how important the airport was to Berks County. Most noticeable is the new control tower built in 1965, and the removal of building 505, replaced by the terminal built in 1961. (Both, courtesy of Terry Sroka.)

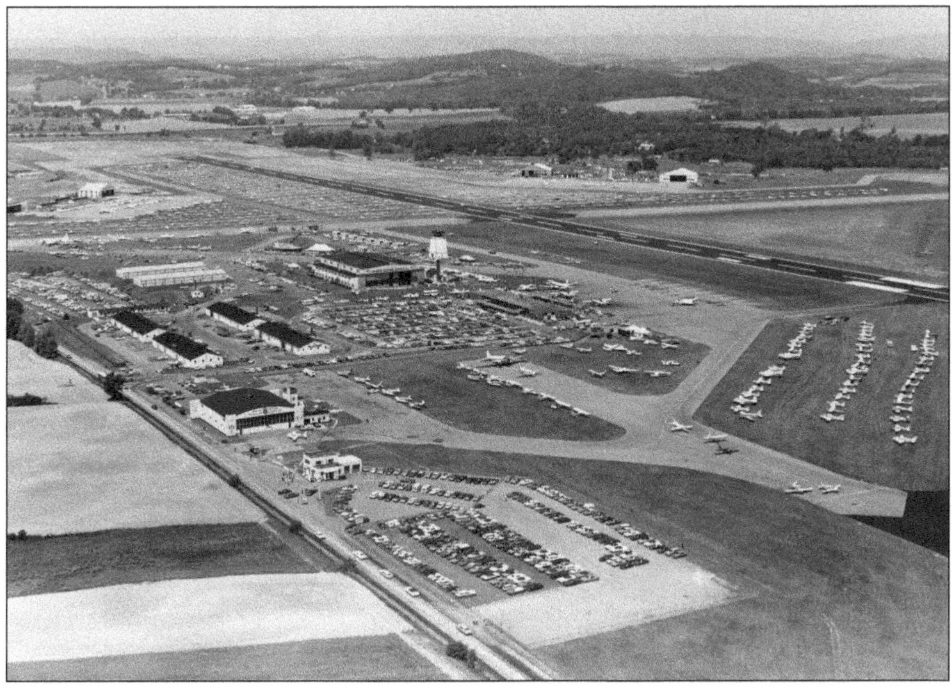

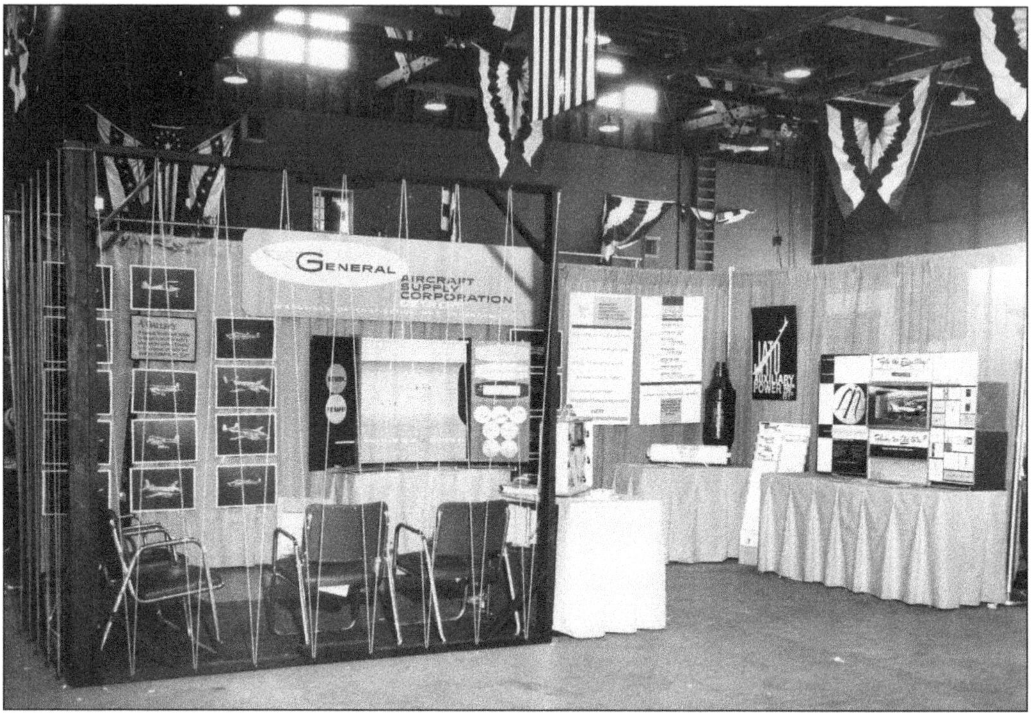

Another example of the increased popularity of the Reading Airshow can be seen in these views of vendor displays. The photograph above was taken in June 1959, and the photograph below was taken in June 1968. As the show grew in size through the years, displays began to include aircraft as well as products. (Both, courtesy of the National Air and Space Museum.)

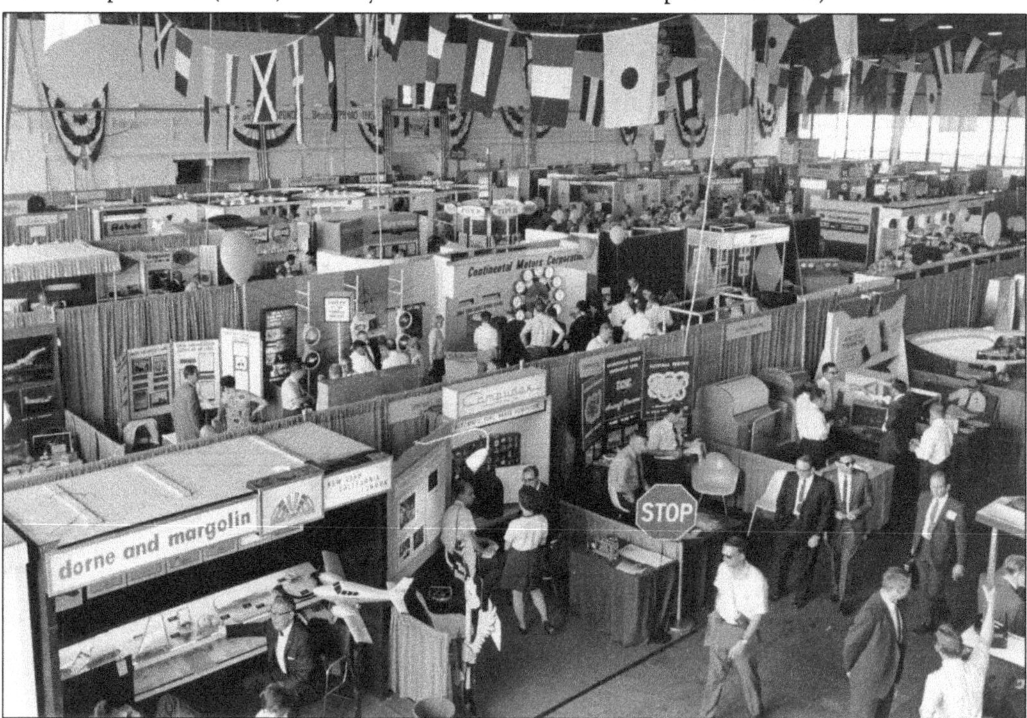

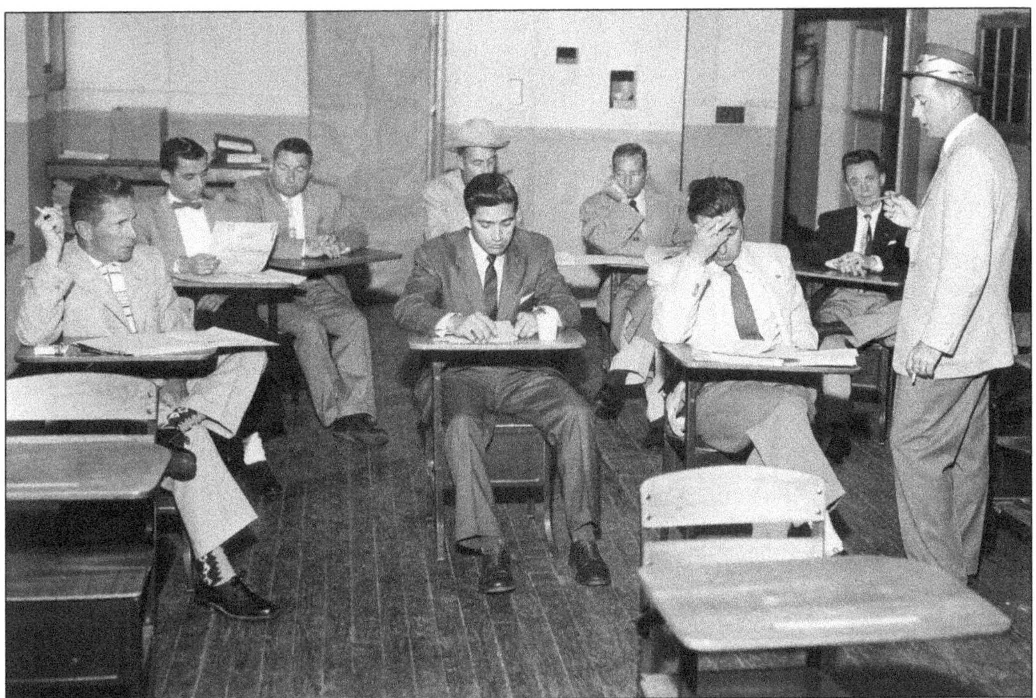

The air show provided competitive and educational events for the national aviation community. The above image shows a meeting of judges for the Reading Aviation Awards competition prior to the start of the Reading Airshow on June 2, 1956. The photograph below shows an air show meeting held during that same event. (Above, courtesy of the National Air and Space Museum; below, courtesy of Terry Sroka.)

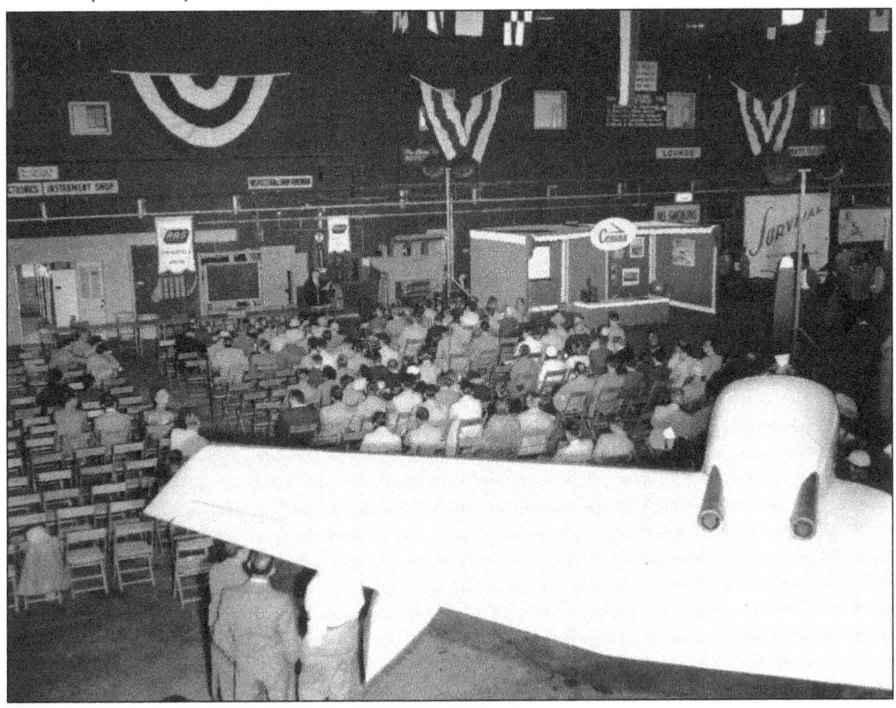

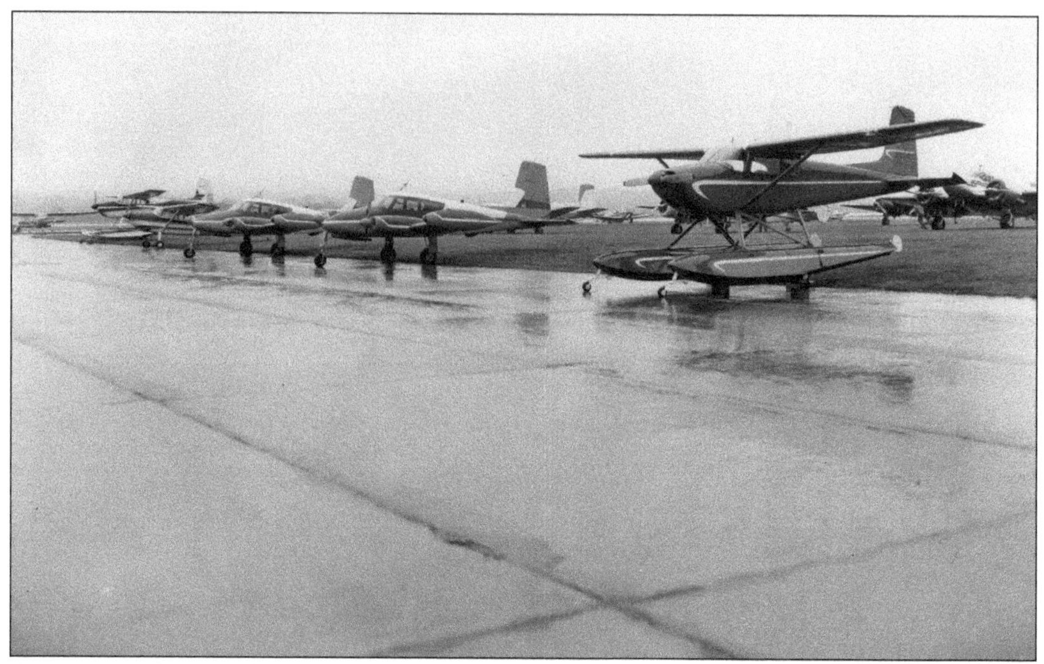

Cessna aircraft are on the ramp at Reading Airport during the Reading Airshow on June 2, 1956. During the 1950s, popularity increased for the annual event, with many commercial and general aviation aircraft lining the ramps on the airport property. This lineup may have been new aircraft on display to potential buyers and dealers. (Courtesy of the National Air and Space Museum.)

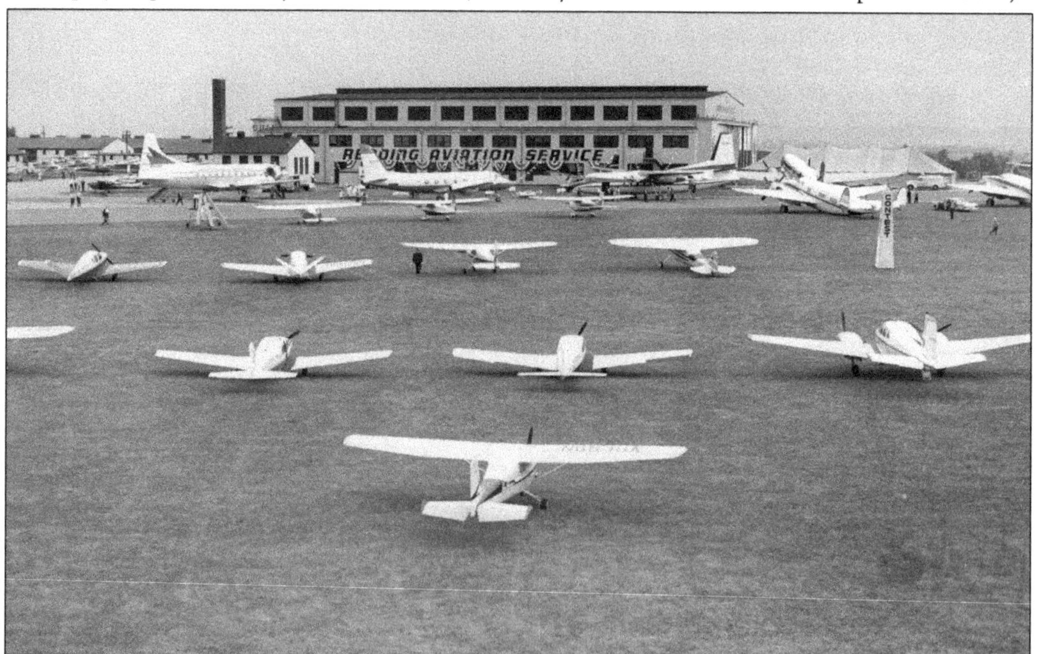

Here, smaller general aviation aircraft are in the foreground ,and larger aircraft are on display in the background closer to the Reading Aviation Service hangar at the Reading Airport during the Reading Airshow in June 1959. The building was also known as Hangar 501 to many airport personnel. (Courtesy of the National Air and Space Museum.)

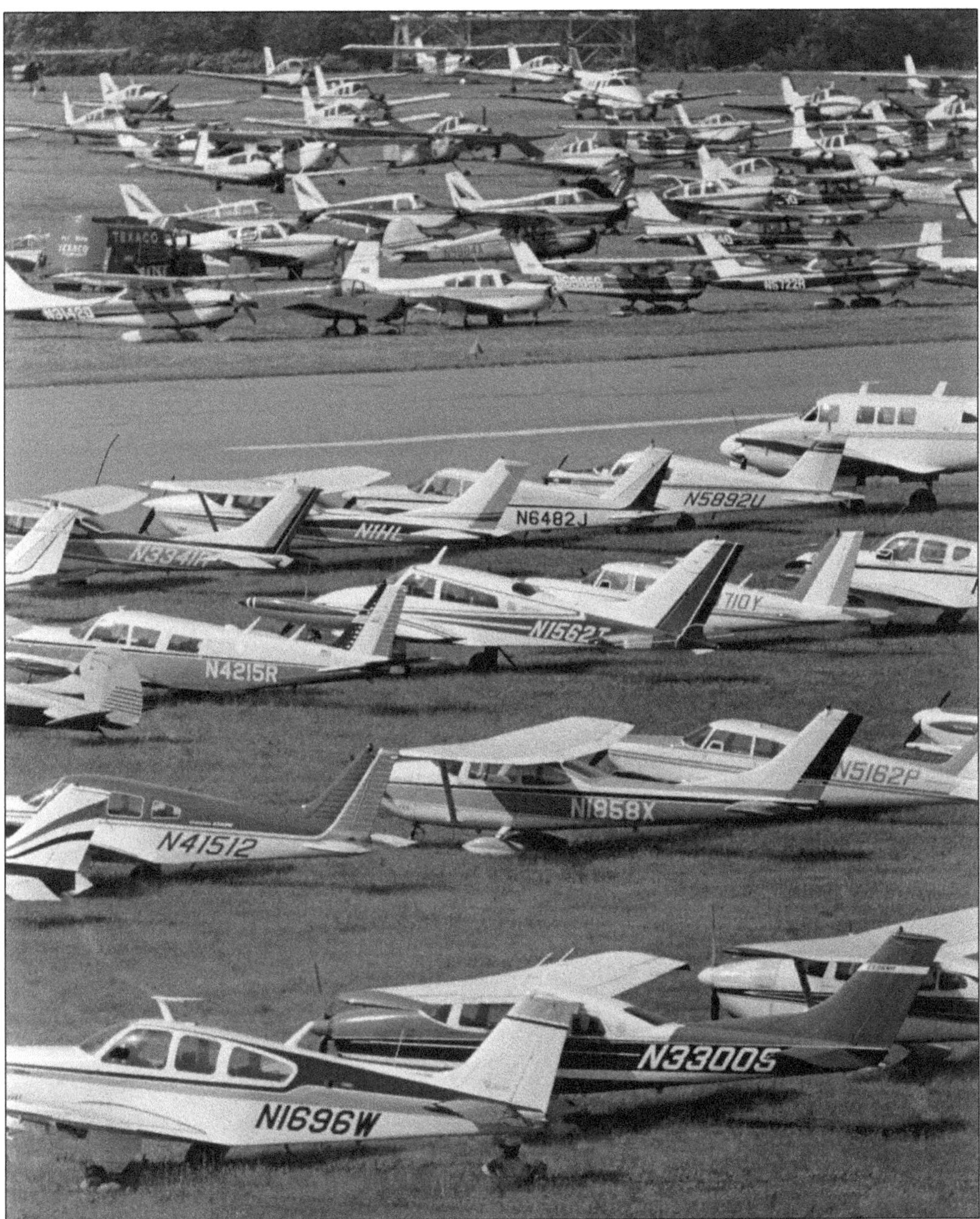

At its peak around 1976, the air show's four-day public attendance exceeded 100,000, with many coming to the show to see the latest aviation products. The air show placed significant burdens on the air-traffic control system. The tower cab housed more than 20 controllers to manage the vast amount of air traffic in the area, with as many as 110 incoming aircraft simultaneously handled by the staff. This photograph, taken at the June 1974 Reading Airshow, shows the volume of aircraft at the event. (Courtesy of the National Air and Space Museum.)

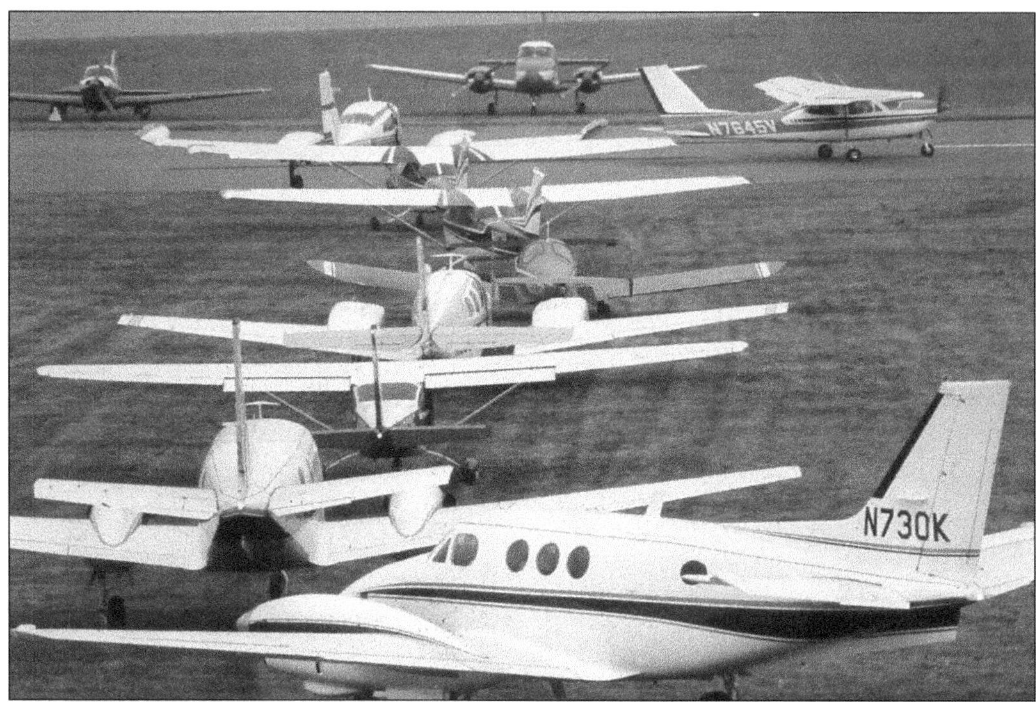

Above are aircraft taxiing after an air show event during the 1970s. Up to 650 airplanes parked on field ramp areas when the air show was most popular. With this influx of activity to the area, local restaurants, bars, and hotels witnessed their greatest business of the year. The image below was taken at the June 1975 Reading Airshow and features the Goodyear *Mayflower* flying above the crowd and a Swearingen SA226-AT Merlin IV. The Swearingen was a twin-turboprop business aircraft that had its maiden flight on April 1965; it was eventually built by Fairchild before production ended in 1987. (Both, courtesy of the National Air and Space Museum.)

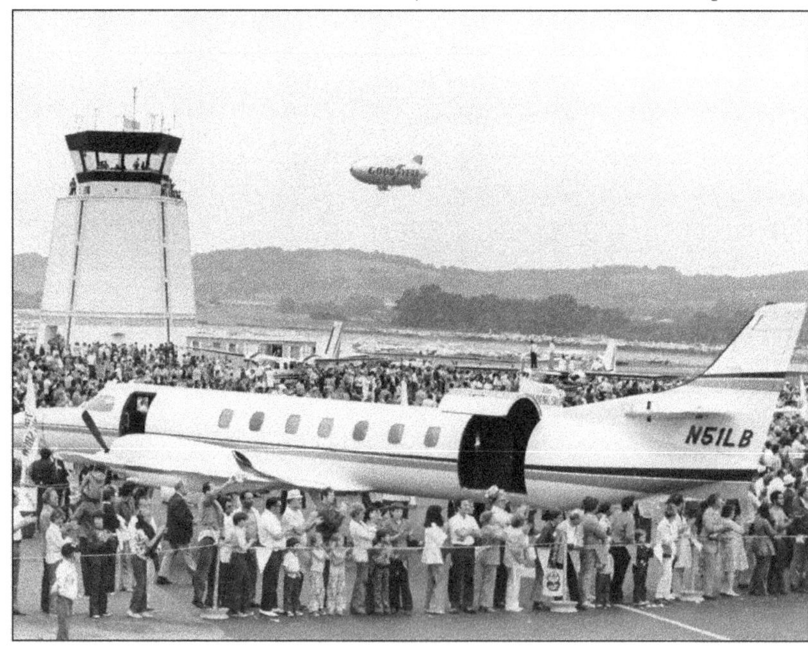

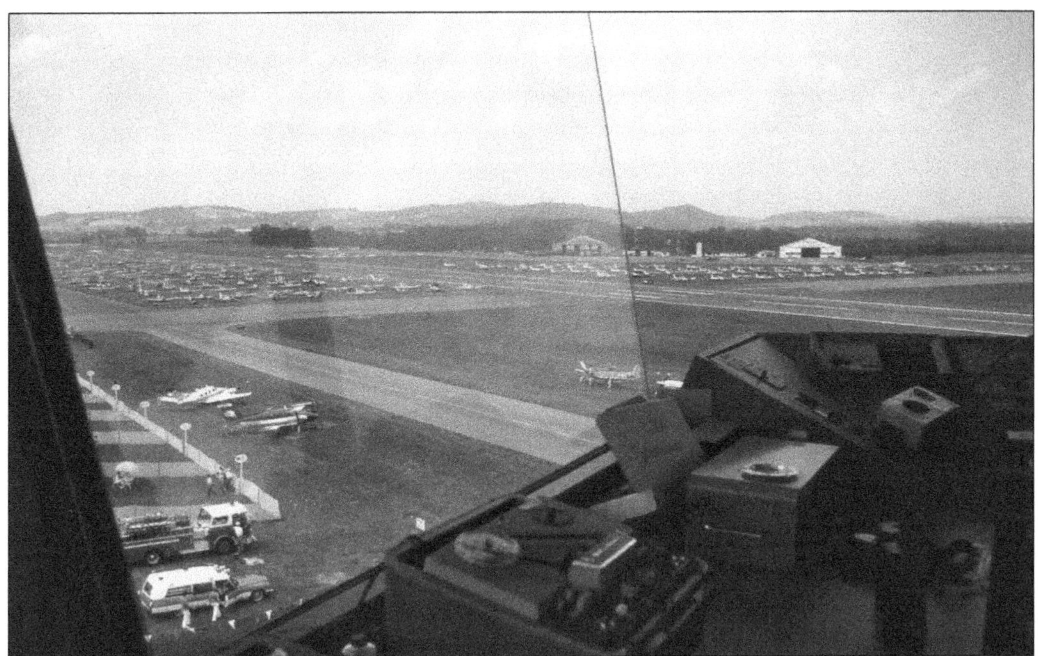

These photographs were taken at the June 1973 Reading Airshow from the control tower, which was built in 1965. The panoramic views show the exceptional visibility air traffic controllers had while managing the flow of aircraft. During the air shows, 20 or more tower personnel worked in the tower cab. In 1966, air traffic was so congested (due to FAA regulations that allowed the whole runway to be reserved for one landing airplane at a time) that air traffic procedures had to be revised. The procedures thereafter eliminated the need to reserve the entire runway for a single arriving or departing aircraft. For the first time, single-engine Cessnas could arrive in trail at 3,000-foot intervals or depart when the aircraft ahead was 2,500 feet down the runway. (Both, courtesy of Terry Sroka.)

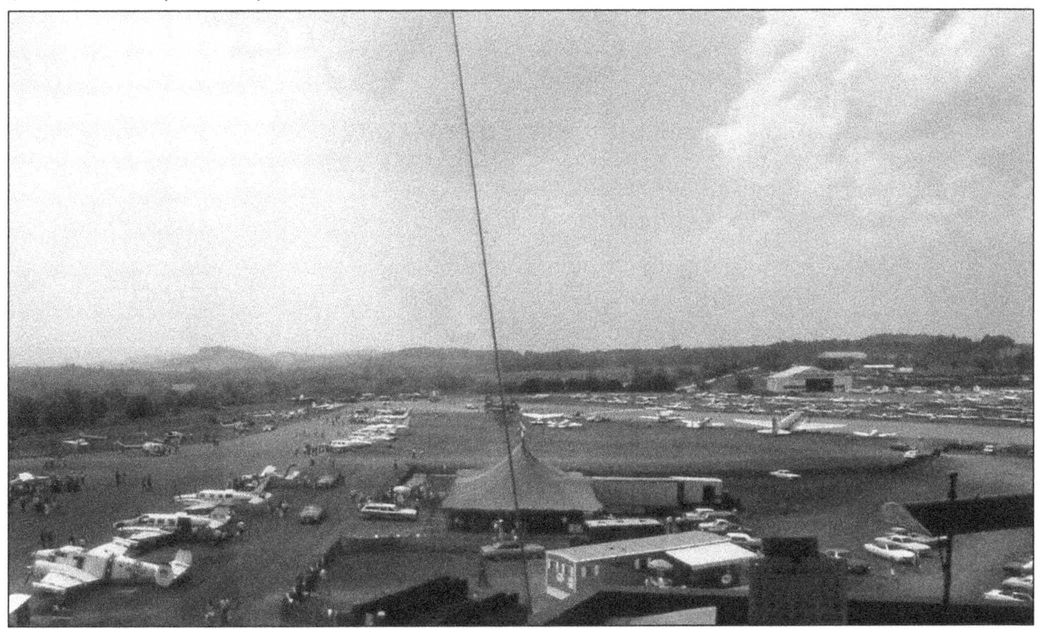

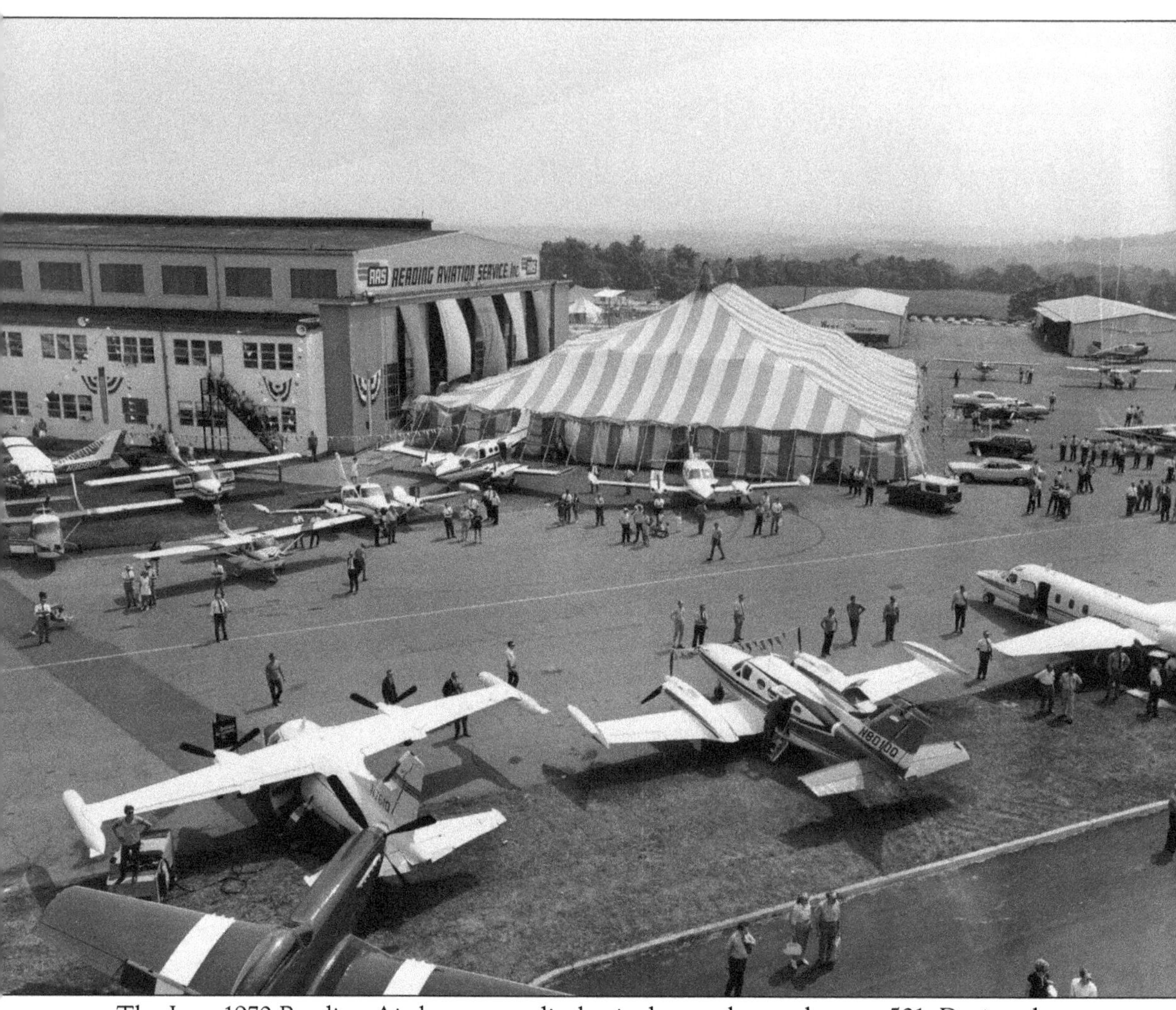

The June 1970 Reading Airshow ramp display is shown close to hangar 501. During that year, the four-day event featured the Air Force Thunderbirds and 157 vendor exhibits displaying aviation products and services. Other educational events were hosted for pilots interested in learning new skills or refreshing their existing aeronautical knowledge. Also, Gov. Raymond Shafer presented the governor's aviation trophy to Alfred Bertolet, founder of Reading Aviation Service, for his outstanding contributions to aviation in Pennsylvania. (Courtesy of the National Air and Space Museum.)

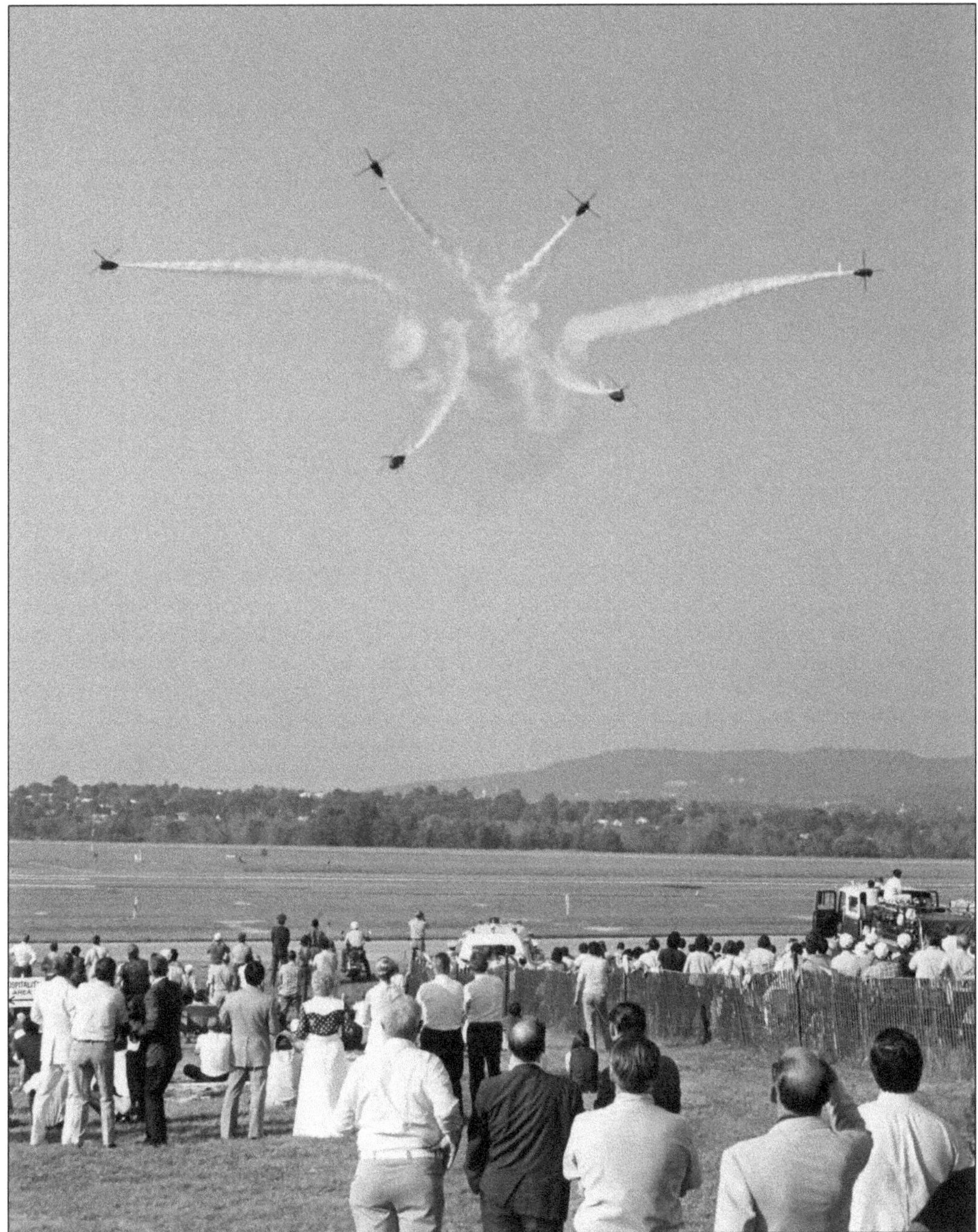

A crowd watches the US Army Silver Eagles aerial demonstration team performing at the Reading Airshow in June 1973. It was not until 1966 that the first military team performed at the Reading Airshow. Since then, the military has been a focal point in the show lineup until a five-year hiatus between 1980 and 1985, when the show was renamed Aerofest, a modern version of the old Reading Airshow. (Courtesy of the National Air and Space Museum.)

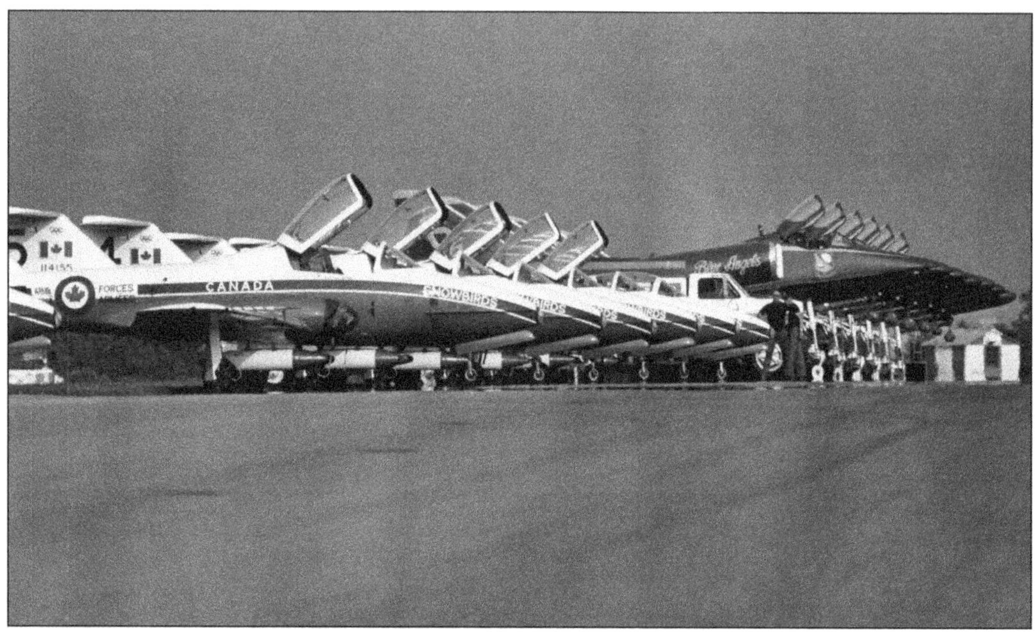

The popularity of the Reading Airshow continued to increase with the introduction of military teams such as the US Air Force Thunderbirds in 1966. Other teams included the Blue Angels, Canadian Snowbirds, and British Royal Air Force Red Arrows. Above are the Snowbirds' CL-41 Tutor and US Navy Blue Angels' Douglas A-4F Skyhawk aircraft lined up along a taxiway during the Reading Airshow in June 1976. Below, three McDonnell Douglas Phantom II F-4s were a highlight at the Reading Airshow. The F-4 Phantom, flown from 1969 to 1974, was also flown by the Thunderbirds during the same period; however, by the mid-1970s, both acts switched to the more economical but underperforming T-38 Talon due to the 1973 oil crisis. Five Talons used the equivalent amount of fuel as one Phantom. (Above, courtesy of the National Air and Space Museum; below, courtesy of Terry Sroka.)

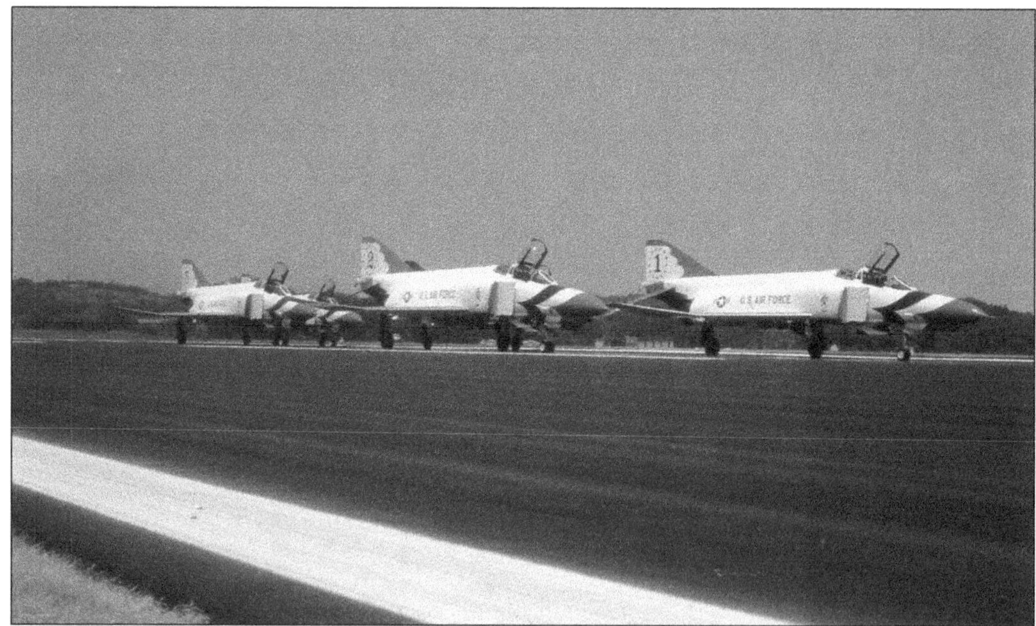

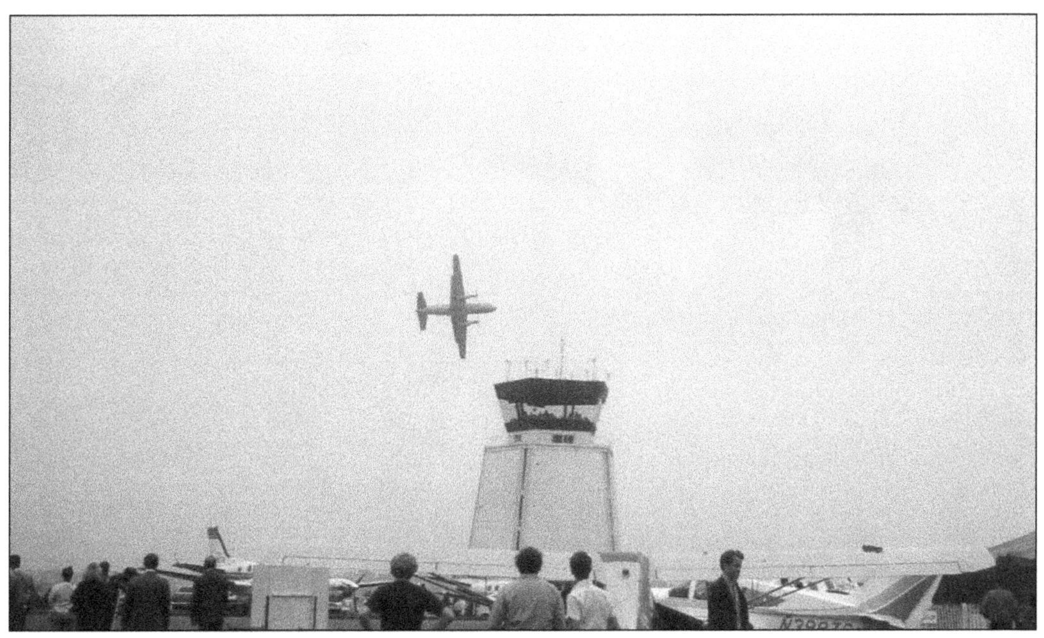

Pictured above and below is the Royal Air Force flying team gaining notoriety with airport controllers and the FAA when they performed an illegal fly-by with their C-130 Hercules, flying within 100 feet of the Reading Airport control tower. (Both, courtesy of Terry Sroka.)

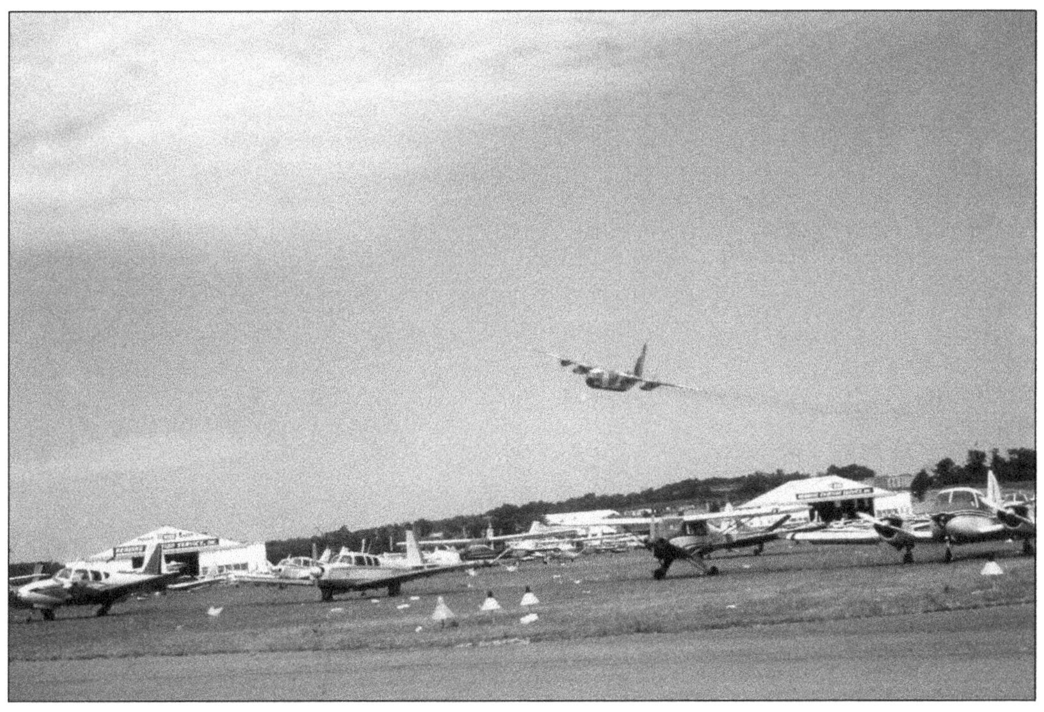

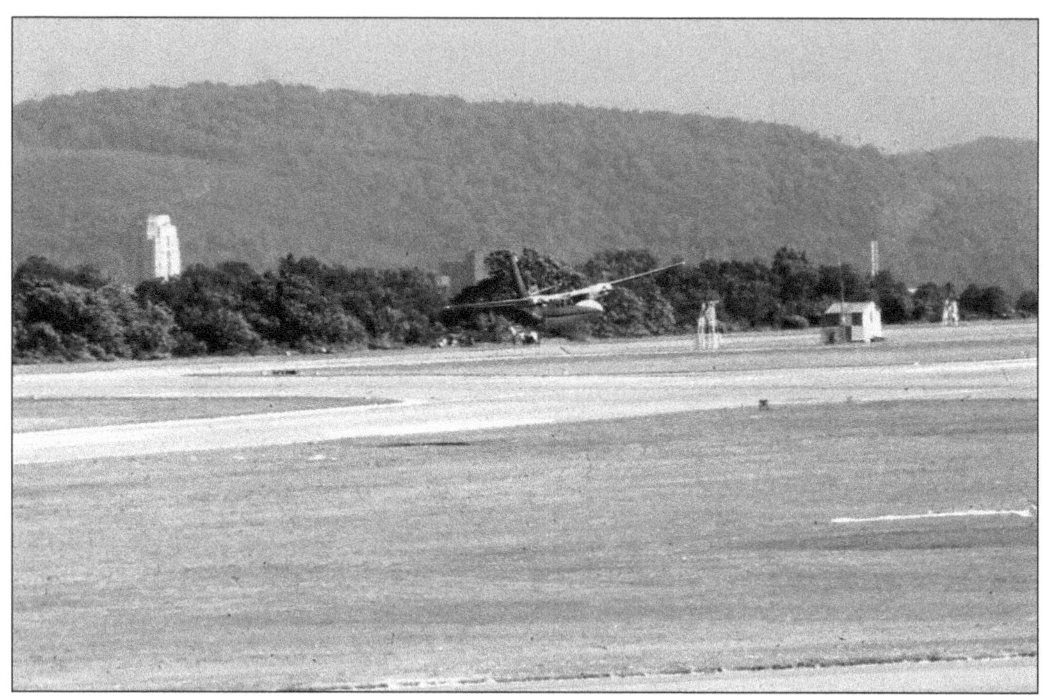

Because Reading Aviation Service was a Rockwell Aero Commander dealer, the 1968 air show was the first time well-known aerobatic pilot Bob Hoover performed in public flying his Rockwell Shrike Commander (above) and P51-D Mustang (below) to promote sales. During Hoover's debut, he scraped the Commander's underside flying too low above the runway surface. In the early 1970s, his P-51 hit a 40,000-volt power line, resulting in a partial blackout of a large section of the city. (Both, courtesy of Terry Sroka.)

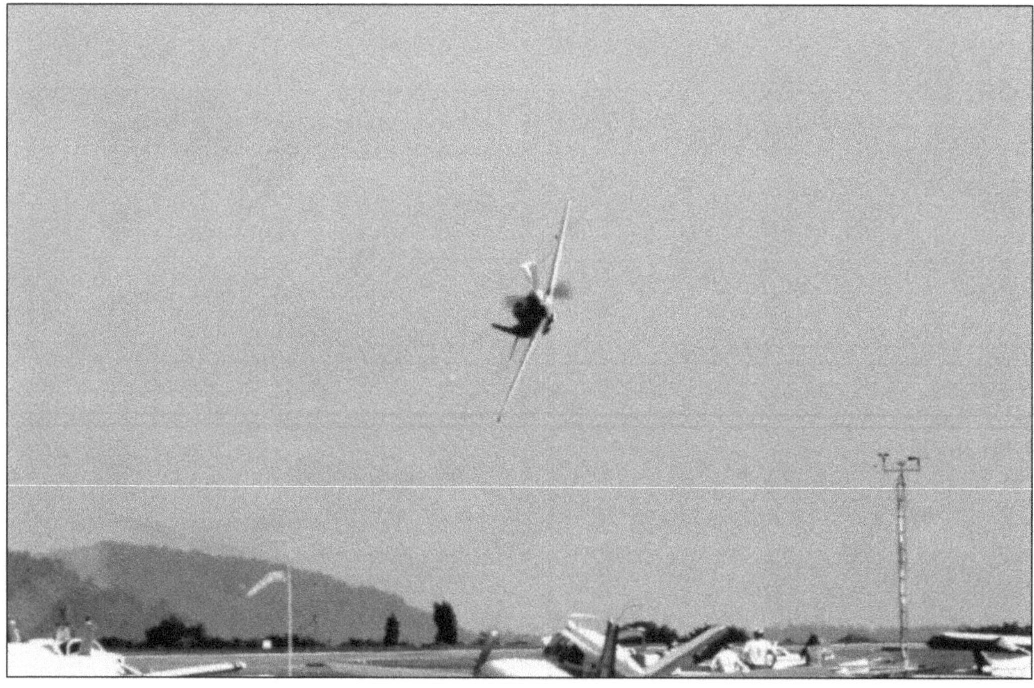

Throughout the years, the Reading Airshow has been a positive event for the county; unfortunately, tragic events were experienced during its extensive history. At the June 4, 1969, air show in front of 10,000 spectators, Capt. Dick Schram, a Naval Reserve aviator known as "the Flying Professor" was killed when his Piper Cub failed to pull out of a dive and crashed within 100 yards of spectators. Another tragic event occurred at the Reading Municipal Airport in 1972 when, on the final night of the show, Reading policeman Charles W. Kurtz killed one fellow officer and wounded another before killing himself during a performance by the Thunderbirds. Witnesses were Kurtz's wife and seven-year-old son. (Right, author's collection; below, courtesy of Blue Angels.)

At Reading Air Show

Policeman Shoots Two Officers, Kills Himself

Special to The Mercury

A Reading policeman wounded two fellow officers before killing himself Friday as they worked as security guards at the Reading Air Show, city police reported.

The dead patrolman was identified as Charles W. Kurtz, Reading. He reportedly shot himself in the chest after wounding the other patrolmen.

Olin J. Wait, Reading, who was shot in the chest and abdomen, was reported in poor condition in the intensive care unit of Reading Hospital late Friday night.

The third policeman, Russell S. Huyett Jr., Shillington, suffered a superficial wound of the chest.

Police reported that Kurtz allegedly began the shooting, firing his service revolver at Wait.

According to investigators, Wait attempted to return the gunfire, but missed Kurtz and dropped to the ground when he was hit by a second bullet.

Huyett, who was near the other two policemen, was reportedly shot by Kurtz when he attempted to take the weapon from Kurtz.

Officials believe that after shooting Wait and Huyett, Kurtz turned his gun on himself and fired the final bullet. The cause of the shooting was not determined.

FINAL NIGHT

The shooting occurred on the final night of the 22nd annual Reading Air Show during the performance of the U.S. Air Force Thunderbirds.

Police said the precision air team was flying through the

(Continued on Page Two)

Sen. Muskie Refuses To Endorse McGovern

WASHINGTON (AP) — Sen. Edmund Muskie said Friday he is not going to hand Sen. George McGovern the Democratic presidential nomination, at least until McGovern makes peace with skeptics in the party.

Conceding that McGovern probably will get the nomination anyway, Muskie said he still was not going to give the endorsement which might make it certain.

In a National Press club speech, Muskie rejected arguments that he should swing his support to the South Dakota senator in order to blunt a "stop McGovern" movement

"Party unity is not achieved with the magic wand of the kingmaker," Muskie said. "No man can hand George McGovern a united party. And I would do him a grave disservice to pretend that I could do so."

Muskie at one time had been considered the Democratic front-runner, but after a series of primary defeats he announced he would end his campaigning in the primaries.

Speculation had mounted this week that he might drop out altogether and endorse McGovern.

Muskie said he was concerned about the new party re-

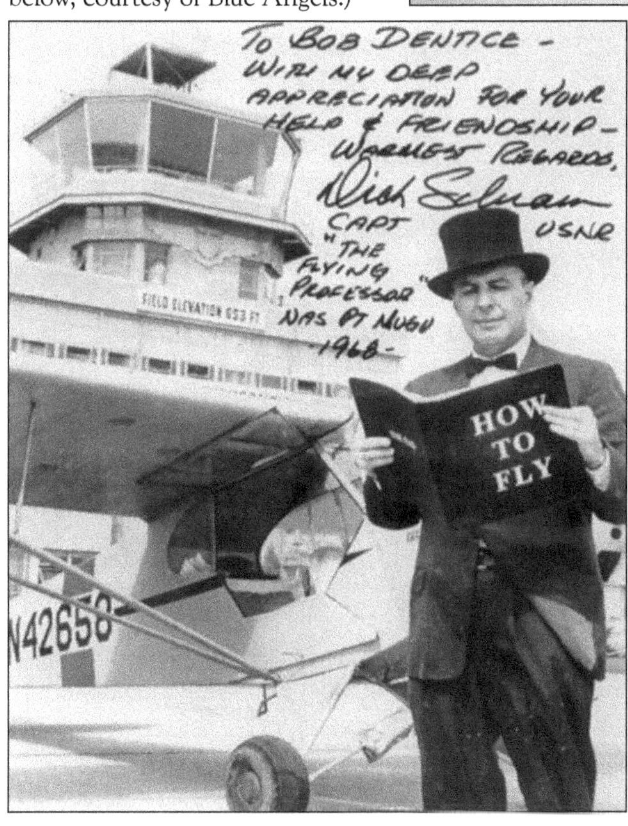

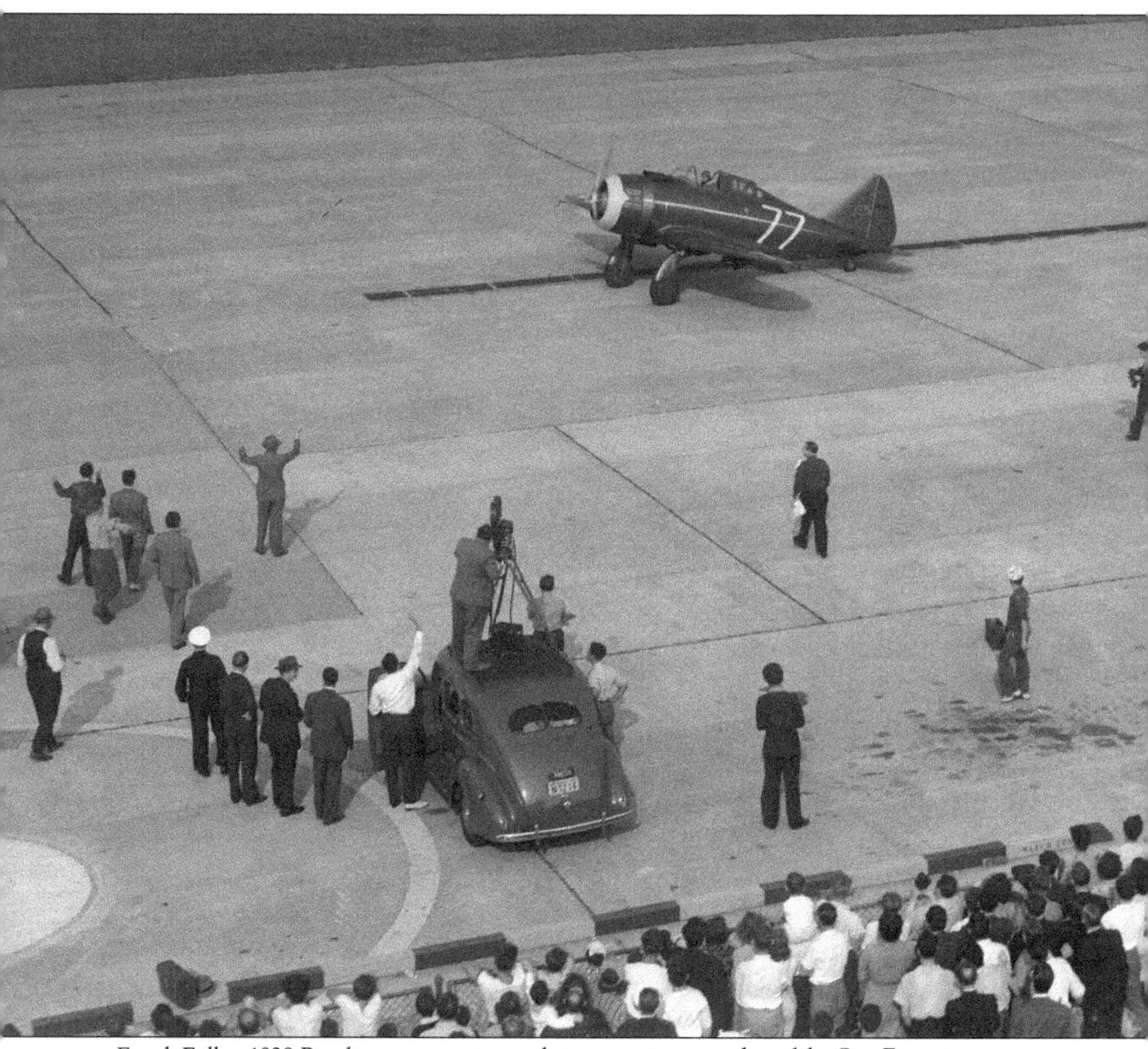

Frank Fuller, 1939 Bendix transcontinental air race winner and wealthy San Francisco sportsman, flew to the new Reading Airport to address the Reading Aviation Association in his Seversky S-2, similar to the one shown here. Flying no higher than 9,000 feet from Cleveland to Reading, he arrived in one hour and eight minutes. More than 100 men and women heard Fuller describe his flight to Reading and the transcontinental air races he won. (Courtesy of the National Air and Space Museum.)

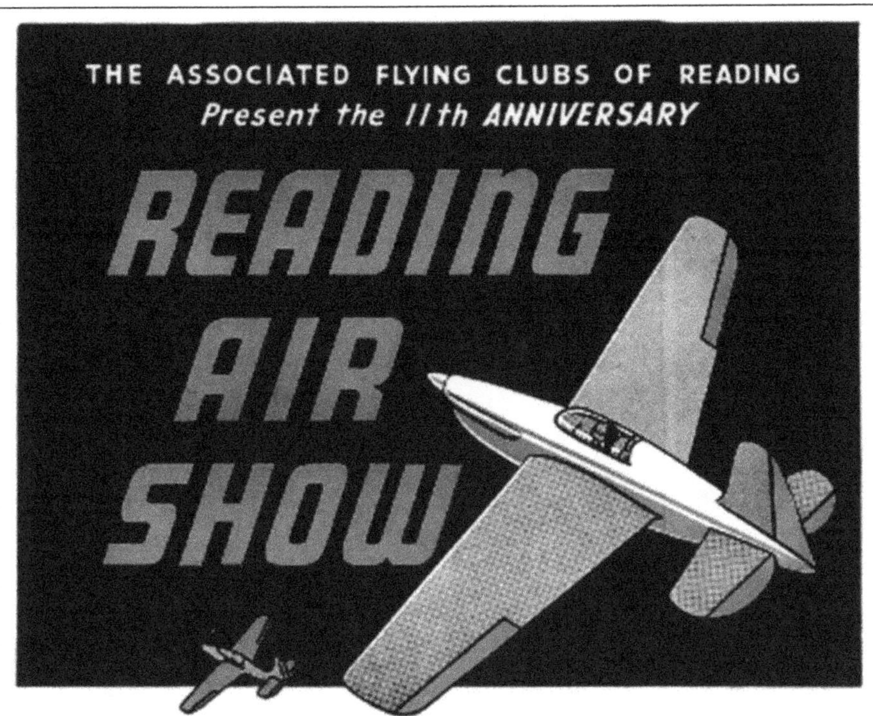

Featuring the

REBAT MIDGET PLANE RACE

SPONSORED BY READING BATTERIES, INC., AVIATION DIVISION

Aerobatics by International Champions

★ ROD JOCELYN ★ BEVO HOWARD ★ BETTY SKELTON

★ JACK HUBER'S SENSATIONAL DELAYED PARACHUTE JUMP

★ COMEDY "PUSHER" RACE ★ WOMEN'S CROSS-COUNTRY RACE

Model Plane Contests 10.30 A. M.

SUN • SEPT • 24 • 1950 • 1 p. m.

READING MUNICIPAL AIRPORT

ADVANCE SALE TICKETS 85¢ • DAY OF RACE $1.00

CHILDREN UNDER 14 ADMITTED FREE • FREE PARKING

In September 1950, the aviation division of the Reading Battery Company sponsored a midget plane race held at the Reading Airport for the second year. The race, with eight participants, involved 20 laps over a six-pylon, two-mile closed course at the airport. The event was part of the 11th annual Reading Airshow, which by then involved numerous aerial entertainment acts. (Courtesy of Terry Sroka.)

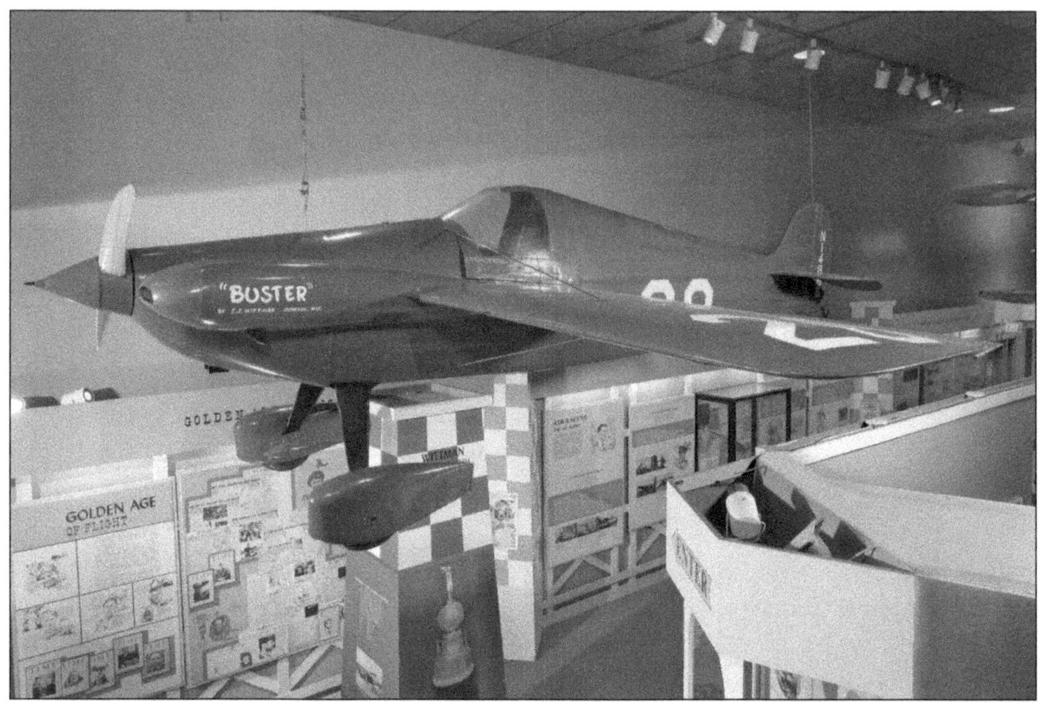

Buster, flown by Bill Brennand, participated in the 11th annual Reading Airshow's air race at the Reading Airport. The Wittman Special 20, powered by a Continental C-85 engine, first flew in 1931 and finished fifth in the air race at 164.446 mph. The airplane is now on display at the Smithsonian Air and Space Museum. (Both, courtesy of the National Air and Space Museum.)

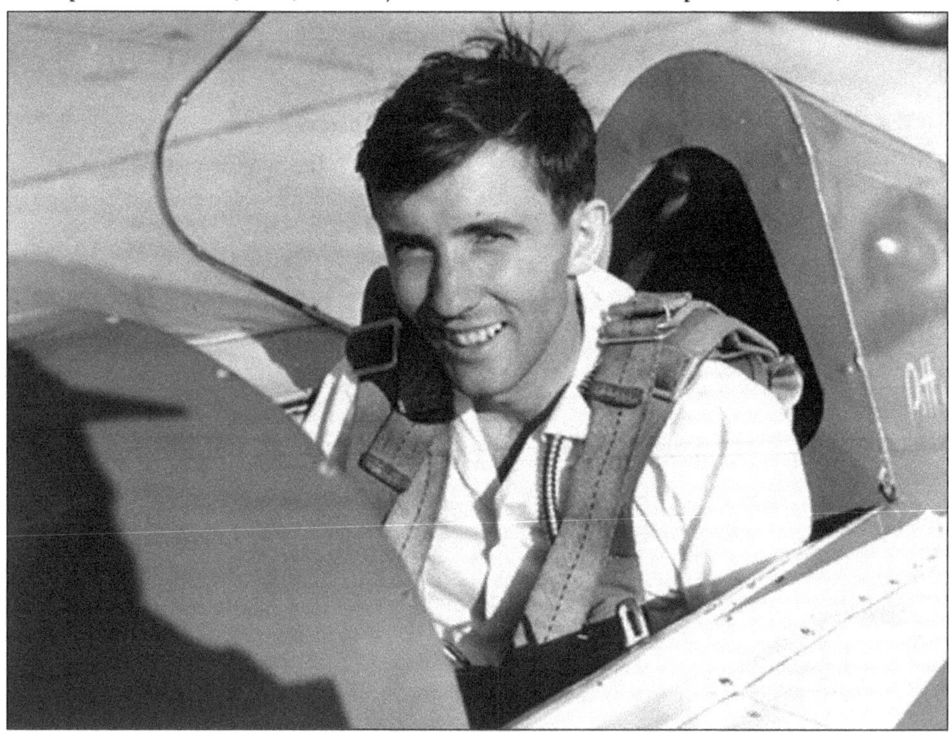

Six

MILITARY AVIATION

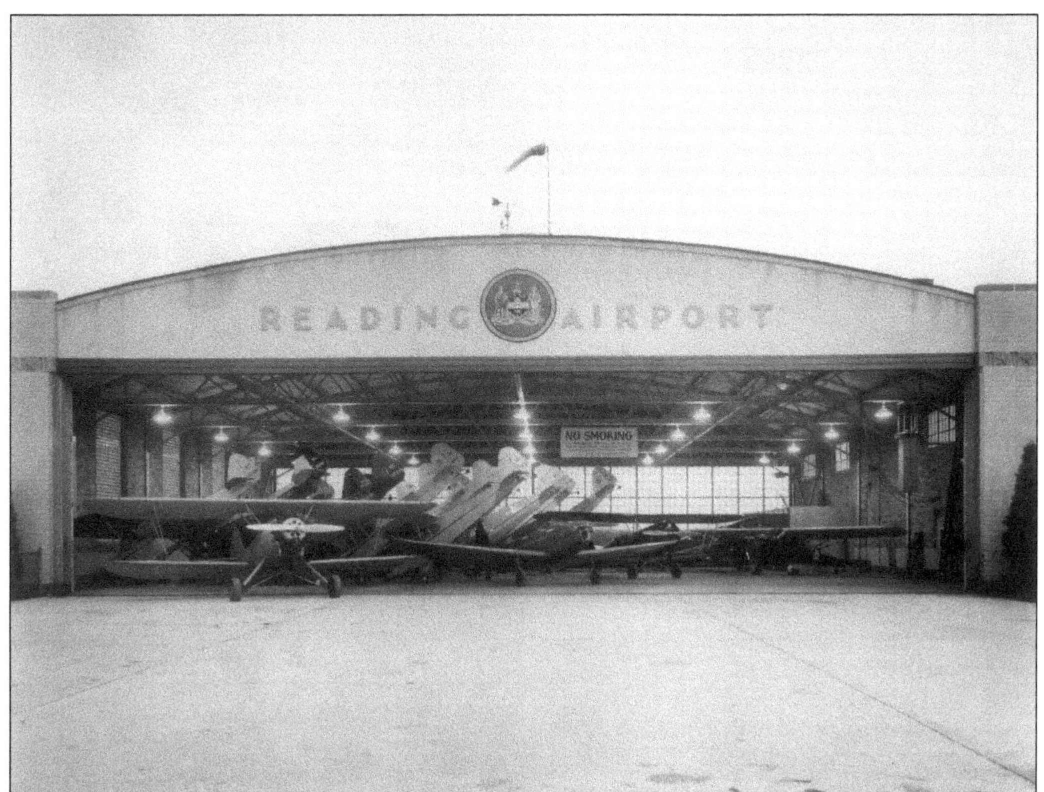

Eight hundred acres of the Reading Municipal Airport were transformed into the Reading Army Airfield on June 1, 1943, which served as a training airfield for tactical reconnaissance units. Shown here are cadet training planes inside the Reading Airport city hangar. Many of the training aircraft used on the field at the time included the Curtiss O-52 Owl, Douglas O-53 and O-46, Stinson O-49 Vigilant, and Aeronca O-58 Grasshopper. (Courtesy of Sime Bertolet.)

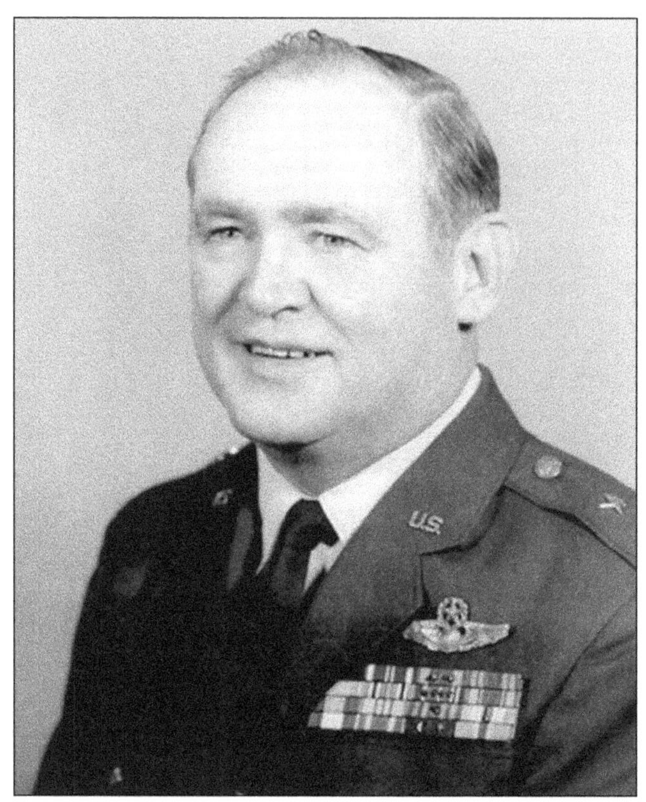

Brig. Gen. Richard Posey, born in Reading, was known as the "Father of the Pennsylvania Air National Guard." In 1942, he served as squadron commander of the 148th Fighter Squadron at Spaatz Field and led an aggressive recruiting effort, recruiting 49 officers and 300 enlisted men. The squadron received federal recognition and began training as a unit of the Pennsylvania Air National Guard. Posey ended several days of prison riots in Bellefonte by buzzing the prison 14 times with his P-51. The prisoners surrendered unconditionally. (Courtesy of the Pennsylvania National Guard Military Museum.)

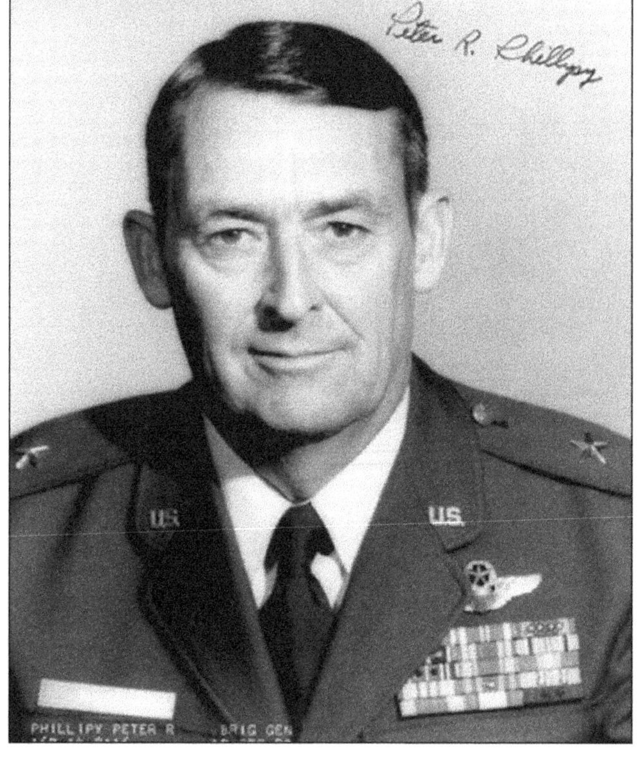

Maj. Peter Phillipy was from Coraopolis, Pennsylvania. On August 2, 1953, he was named commander of the 148th Fighter Squadron. By the mid-1950s, he was promoted to base commander at General Spaatz Field and performed a dual role as base commander of the 112th Fighter Interceptor Group (146th and 147th at Pittsburgh and the 148th at Reading). The 148th was given that designation because of its geographic location. (Courtesy of the Pennsylvania National Guard Military Museum.)

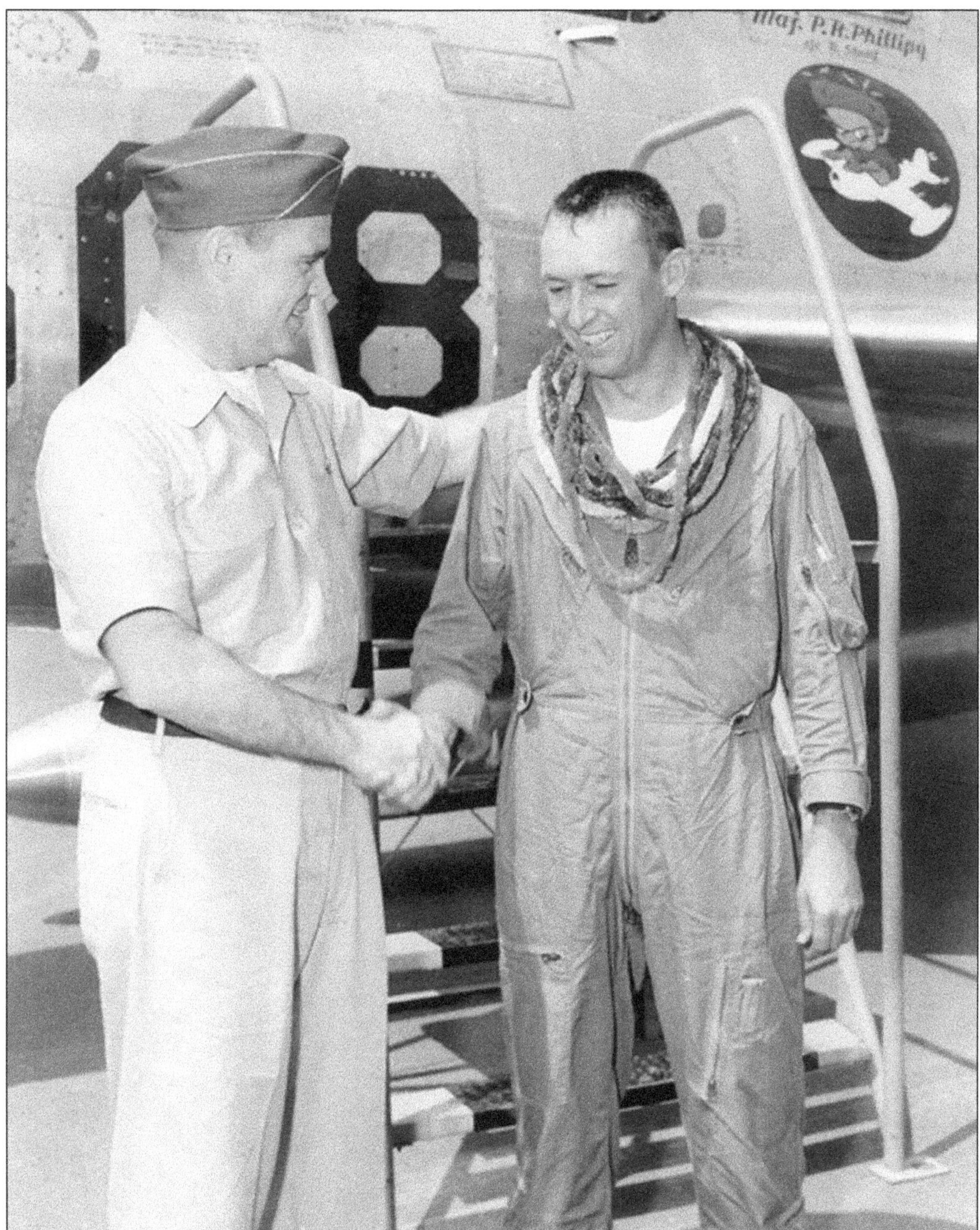

Col. Richard Posey (left), commander of the 112th Fighter Interceptor Wing, Pennsylvania Air National Guard, congratulates Maj. Peter Phillipy upon his return from Handcock Field in Syracuse, New York, after winning the Ricks Memorial Trophy in 1957. The race was from Fresno, California, to Washington, DC, and Phillipy set a new world record flying an F-84F at 638 mph in 4 hours and 13 minutes. This was an Air Force–sponsored cross-country jet race by Air National Guard aircraft and crews named after Maj. Earl T. Ricks, who died in January 1954. (Courtesy of the Pennsylvania National Guard Military Museum.)

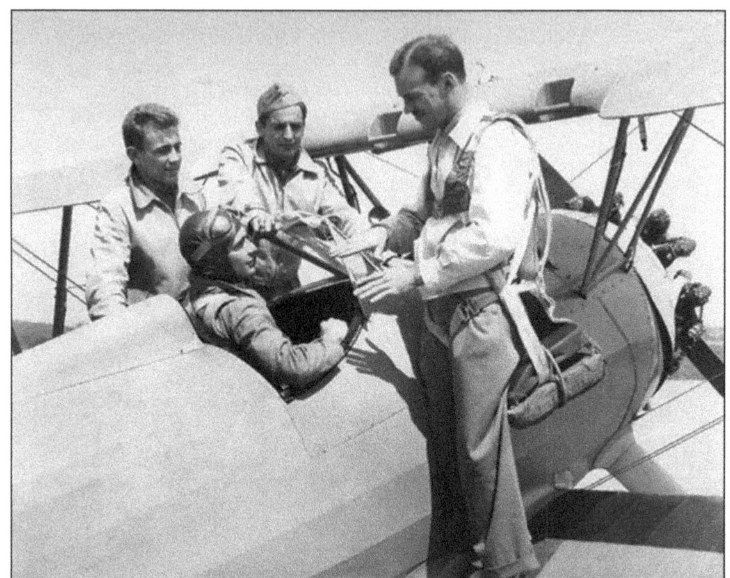

The 7th College Training Detachment of the Albright College cadet program prepared students to become pilots. Some of the men had never been up in an airplane, while others already had their civilian pilot licenses. The training included up to 10 hours of instruction, with many never going on to become military pilots after the program ended. (Courtesy of Albright College.)

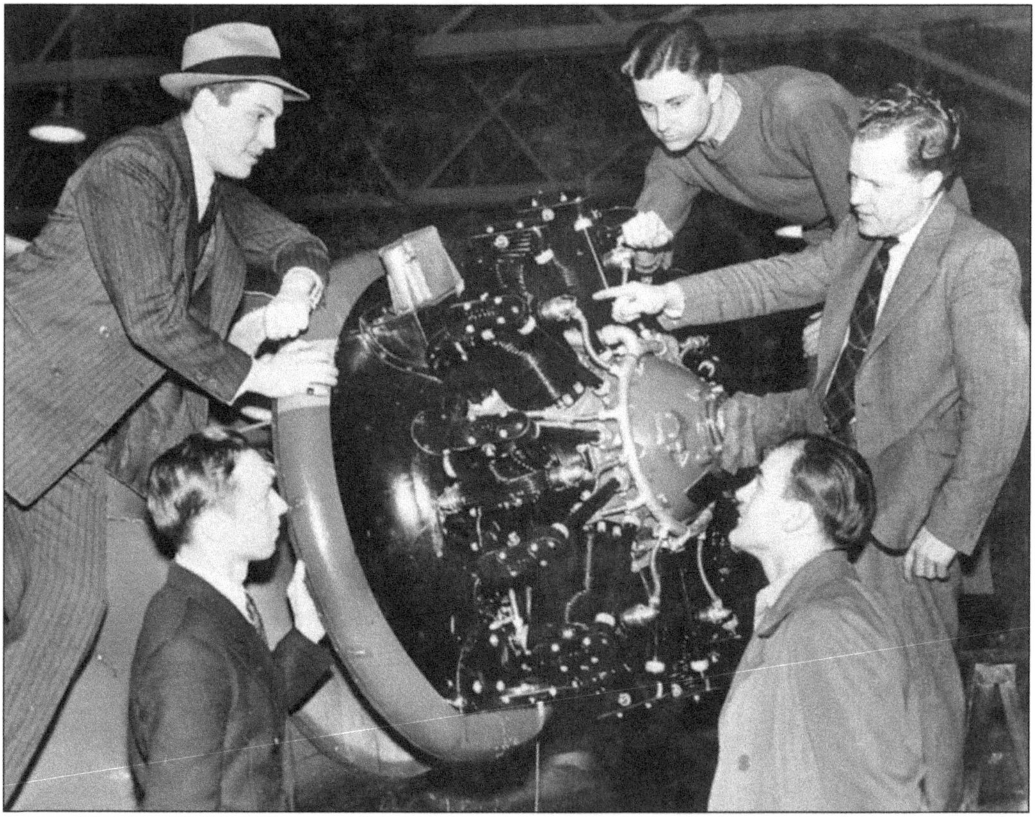

Part of the 7th College Training Detachment of the Albright College cadet program included coursework in the mechanical operations of aircraft. These courses prepared students to not only become pilots, but also to become navigators or bombardiers to help the war effort. The programs were accelerated by reducing summer breaks and offering 12-week summer sessions. (Courtesy of Albright College.)

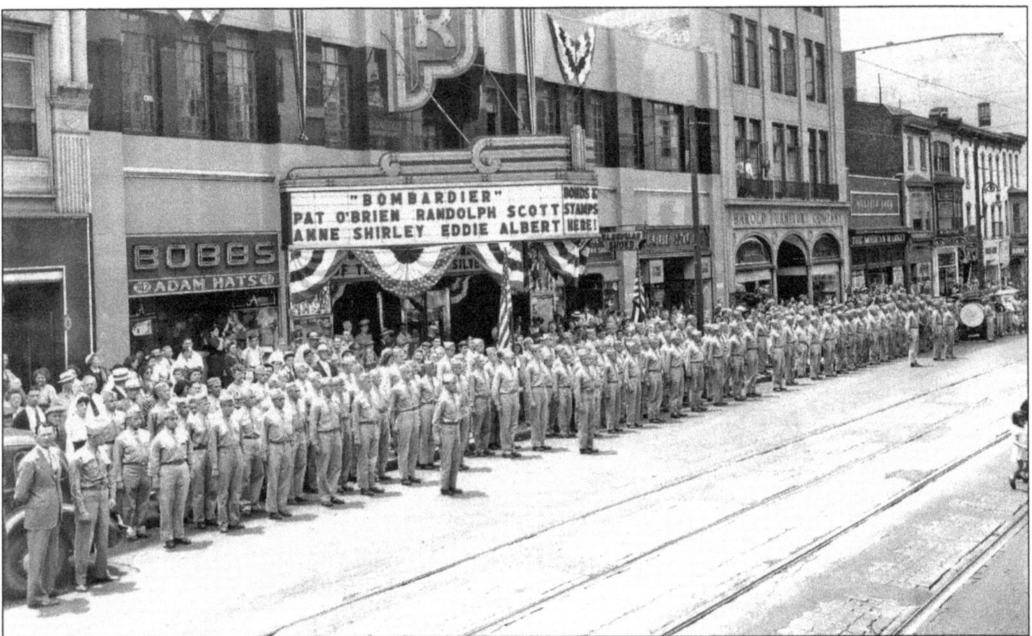

In the fall of 1939, the Civil Aeronautics Authority sponsored an aviation school for Albright College students. These students were known as the "Flying Lions." In 1942, fifty-two air cadets began an eight-week course on the college campus. By March 1943, there were 328 men and women receiving military training. Many of these cadets are shown standing along Penn Street in front of the Astor Theatre. Some cadets went on to the military and achieved high honors, such as Fleetwood's Paul Schlegel. Pictured below on the USS *Enterprise* in 1942, Schlegel (third row, second from right) received the Navy Cross and two air medals for fighting in the Pacific during World War II. He went on to become a Navy commander. (Above, courtesy of Albright College; below, courtesy of Robert J. Cressman.)

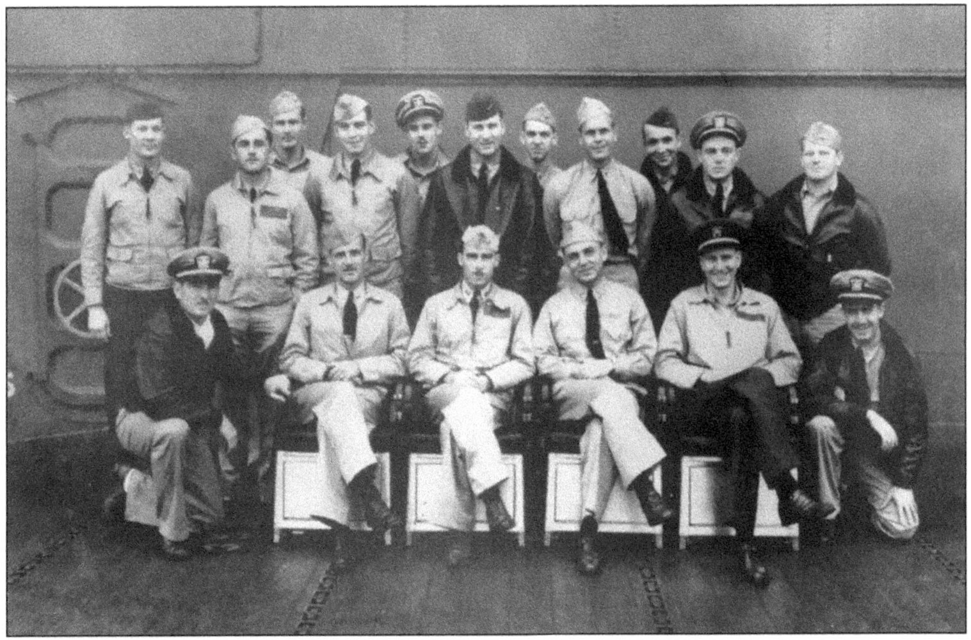

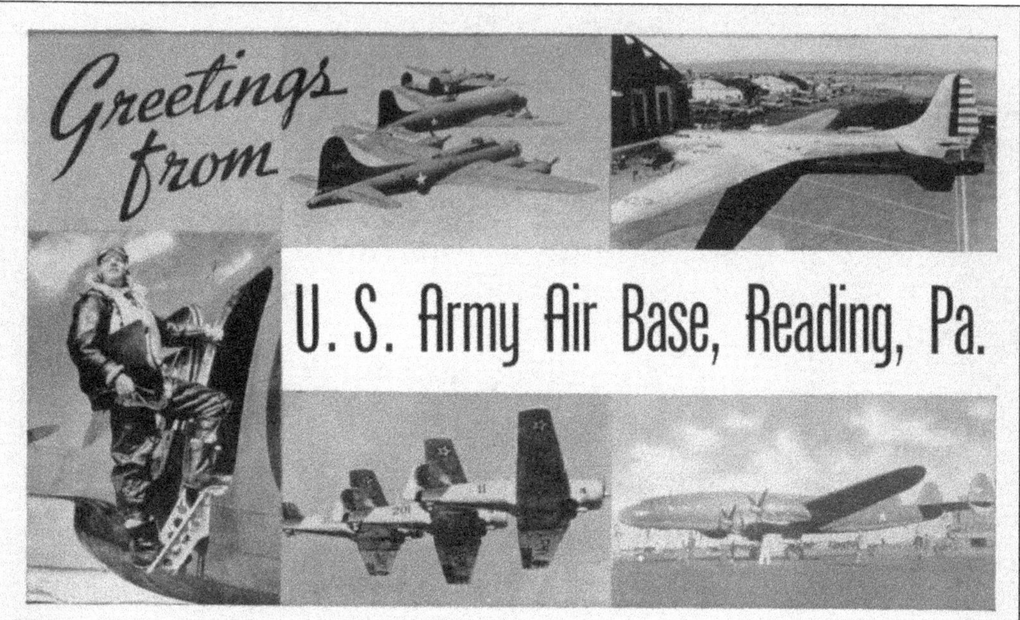

Most likely, those who were stationed at the Reading Army Air Base sent postcards like this to family and friends during the 1940s. The promotional scenes on this card were not necessarily photographed at the airfield; however, the postcard is a rare and attractive military marketing document nonetheless. (Author's collection.)

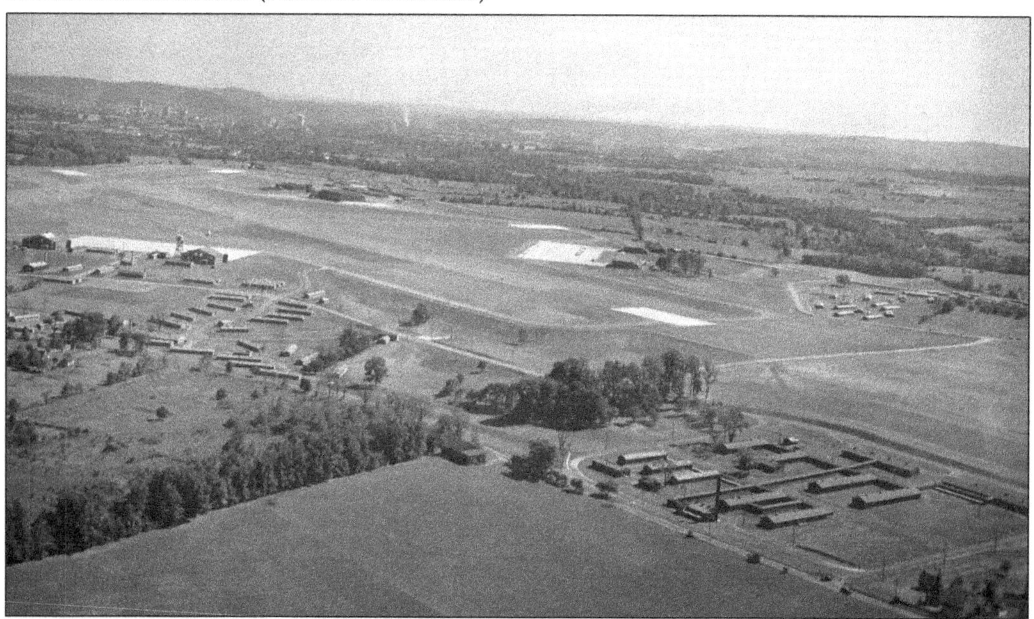

Reading Army Airfield opened on June 1, 1943, with the 309th Base Headquarters and Air Base Squadron as its host unit. Its mission was to train tactical reconnaissance units. The 26th Tactical Reconnaissance Group was activated on the airfield on the same date, with the 37th, 39th, 40th, and 91st Photo Reconnaissance Squadrons. This photograph, taken in 1943, shows the northern side of the airfield with barracks used for personnel, and eventually, German prisoners of war. (Courtesy of Terry Sroka.)

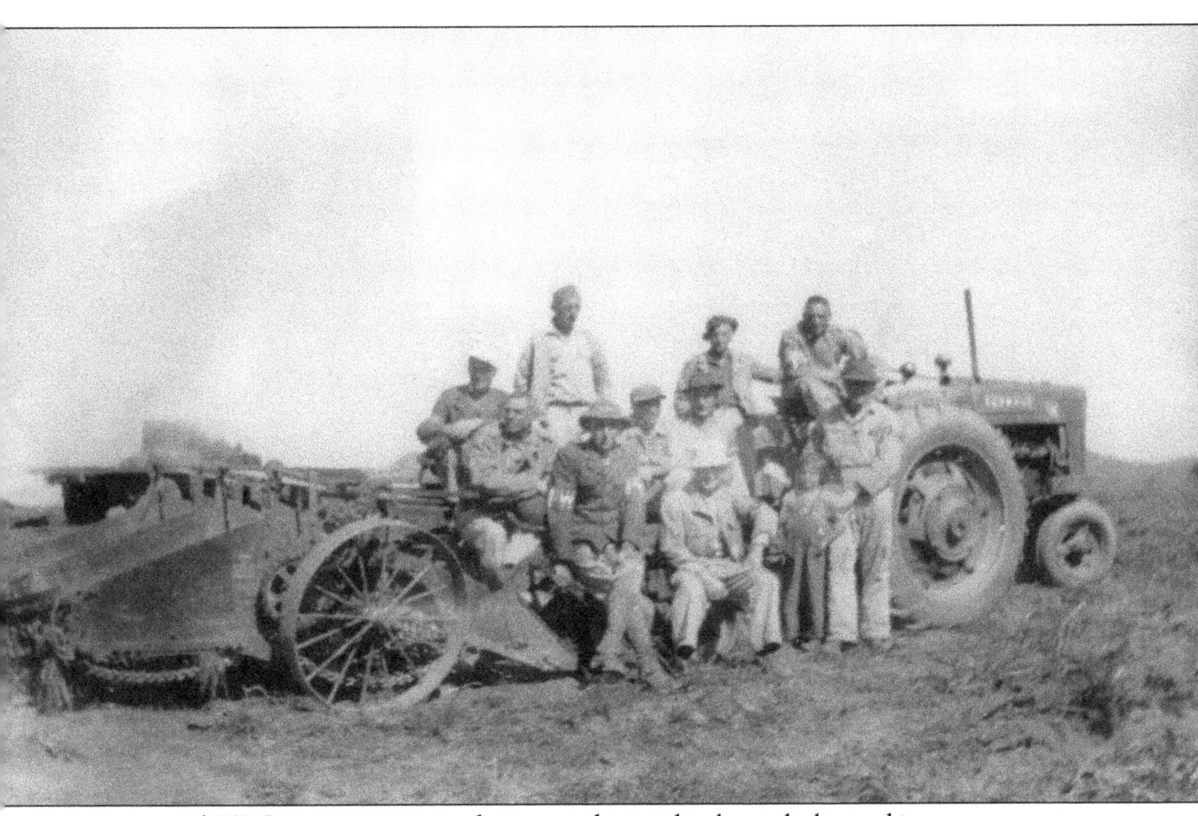

An estimated 250 German prisoners of war were detained in barracks located in a remote sector on the north side of the Reading Army Air Field during World War II. Residents living across the river from these barracks would hear the Germans singing at night. Military police were assigned to guard the prisoners in the camp and while they worked at local farms. By September 26, 1944, local orchards bused the prisoners from the airport stockade to their groves, where the prisoners picked fruit, mainly apples. They wore "PW," short for prisoner of war, on their trousers and shirts. In this undated photograph, German war prisoners pose on the Henry Eyrich farm near Yellow House. After the war, some German POWs became citizens of the United States and lived in Berks County. (Courtesy of the Oley Valley Community Library.)

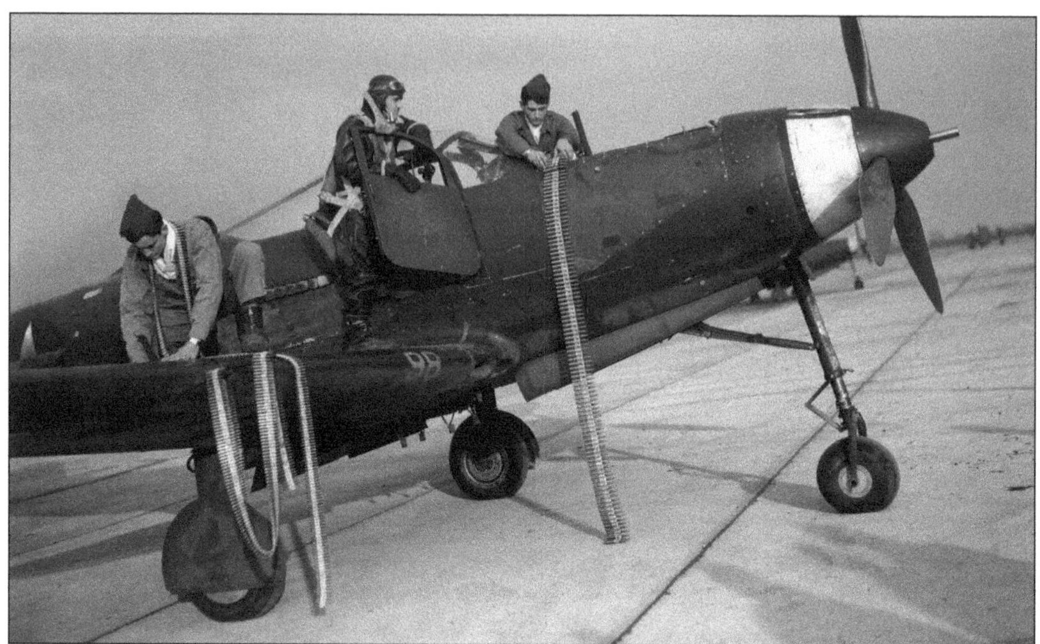

Bell P-39 Airacobras, similar to the one shown here, were found at the Reading Army Airfield. They were prepared for combat before being shipped overseas. A July 25, 1943, edition of the *Reading Times* featured Airacobra 29722 shooting at a mound of dirt on the airfield while harmonizing its guns and propeller. To anchor planes down for the shooting, strong tie-downs were required, which still can be found at the airport today. Signage found on the airfield stated, "Aircraft and guns must be securely fixed in immovable position before firing." (Courtesy of US Air Force.)

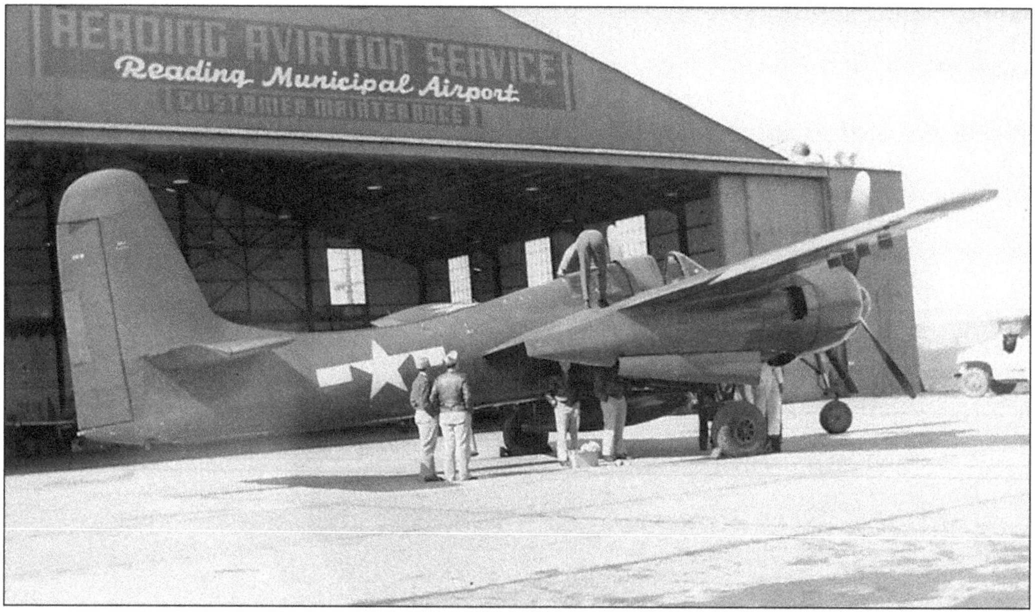

Pictured in front of the Reading Aviation Service customer maintenance hangar in 1946 is an unidentified World War II aircraft. Many of these planes resided on the airfield, but as the Pennsylvania National Guard increased its presence, they were replaced by other types of planes during the 1950s. (Courtesy of Terry Sroka.)

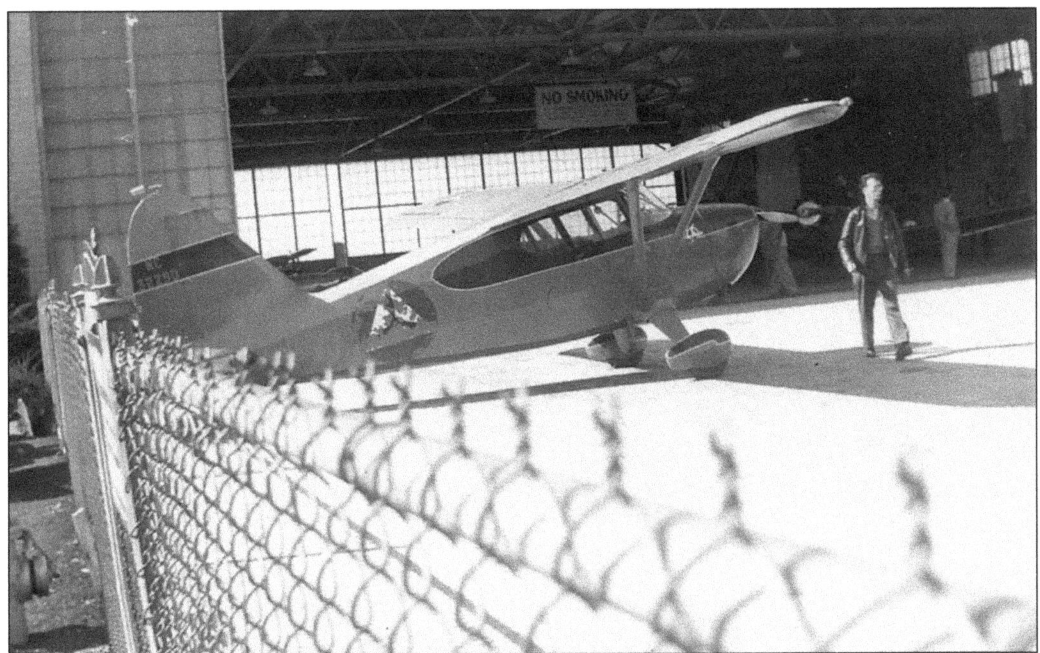

These pictures show Reading Civil Air Patrol Pipers during the 1940s. Above, the Civil Air Patrol emblem is seen, but worn for aircraft NC32290 registered under the Civil Aeronautics Association. Below, brothers Lt. Ralph Burns (left) and Lt. Robert Burns help Lt. Mary McGowin put on her parachute in 1943. (Above, courtesy of Terry Sroka; below, courtesy of George Brizek.)

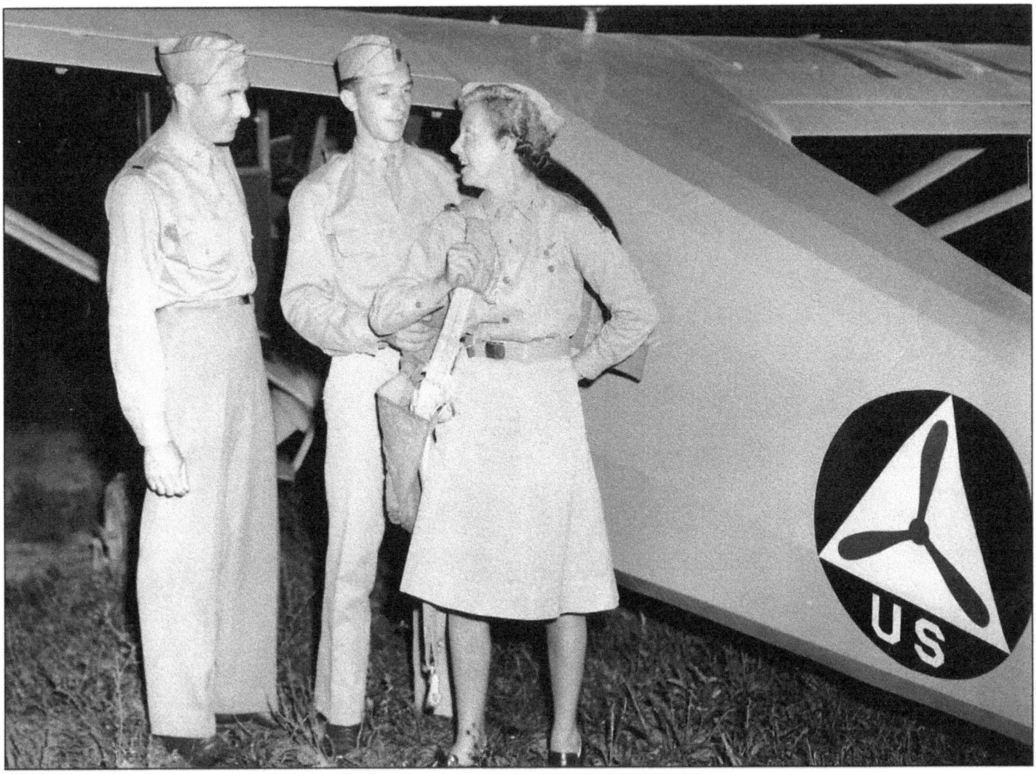

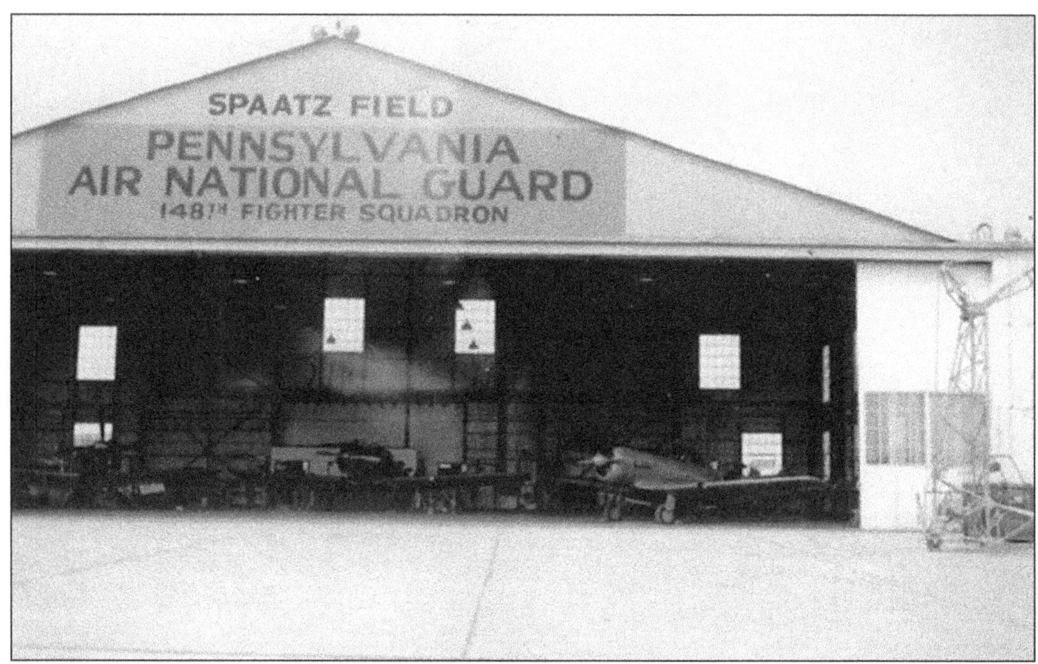

The 12th Air Force 347th Fighter Squadron, reallocated as the 148th Fighter Squadron in May 1946, was an international unit in Reading and was mostly equipped with Republic F-47 Thunderbolts (above) and P51 Mustangs (below left and next page). Later, the F-51D was replaced by jet-propelled Lockheed F-94 Starfires and F-86 Sabres (below), but because the Reading Airport Commission and the National Guard were in conflict on the use of the airport for tactical jet operations, the Air Force inactivated the squadron on June 30, 1956. (Above, courtesy of Terry Sroka; below, courtesy of Cyndi Thomas Deerfield.)

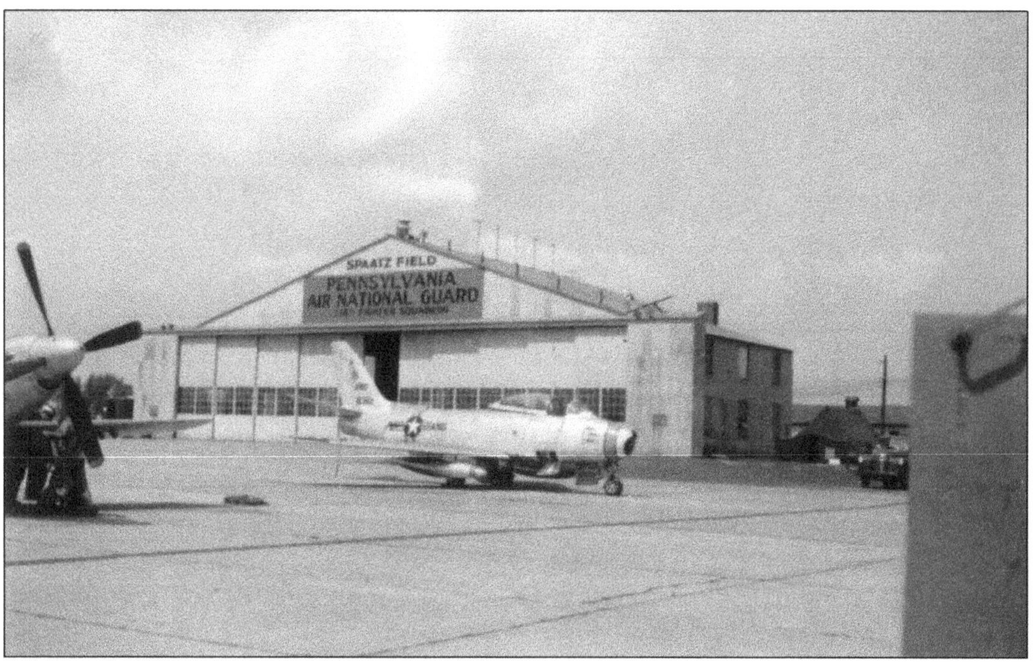

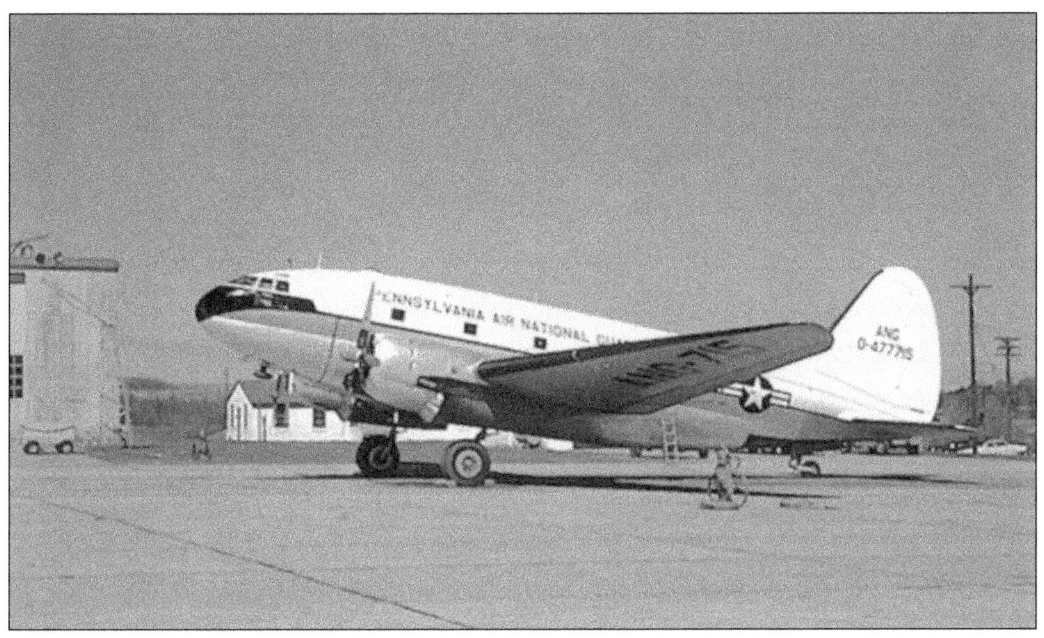

Some of the aircraft used by the Pennsylvania Air National Guard included the Curtiss C-46D seen above as part of the 140th Aeromedical Transport Squadron, or the F-51D Mustang of the 148th Fighter Squadron seen below, both pictured at Spaatz Field in 1957. The C-46D was used as a military transport during World War II by the US Army Air Forces, Navy, and Marine Corps. The F-51D was used as a fighter and fighter-bomber during World War II and the Korean War. (Both, courtesy of Terry Sroka.)

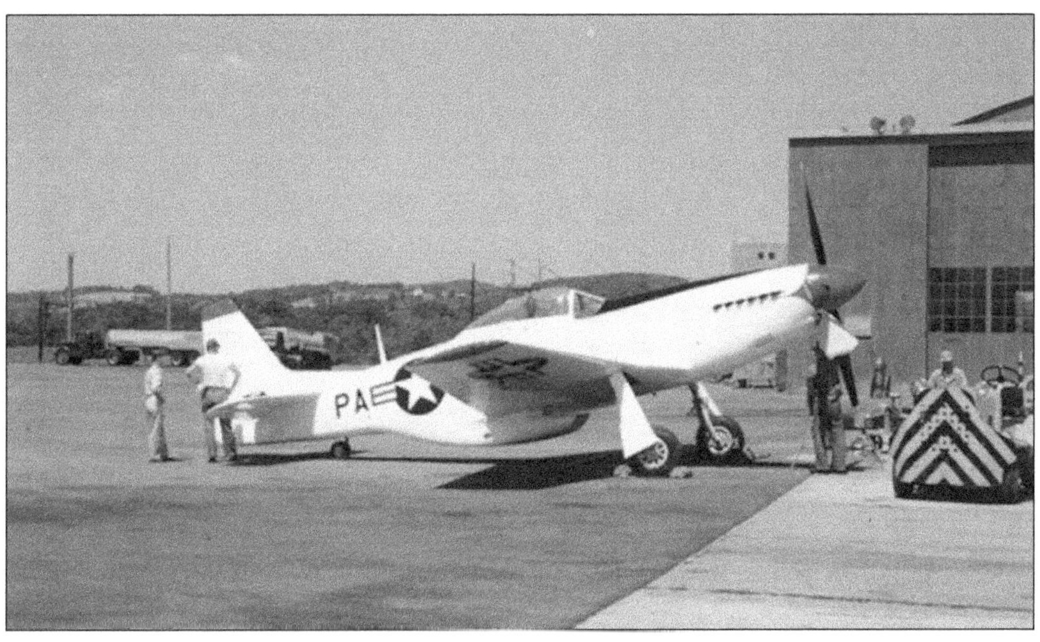

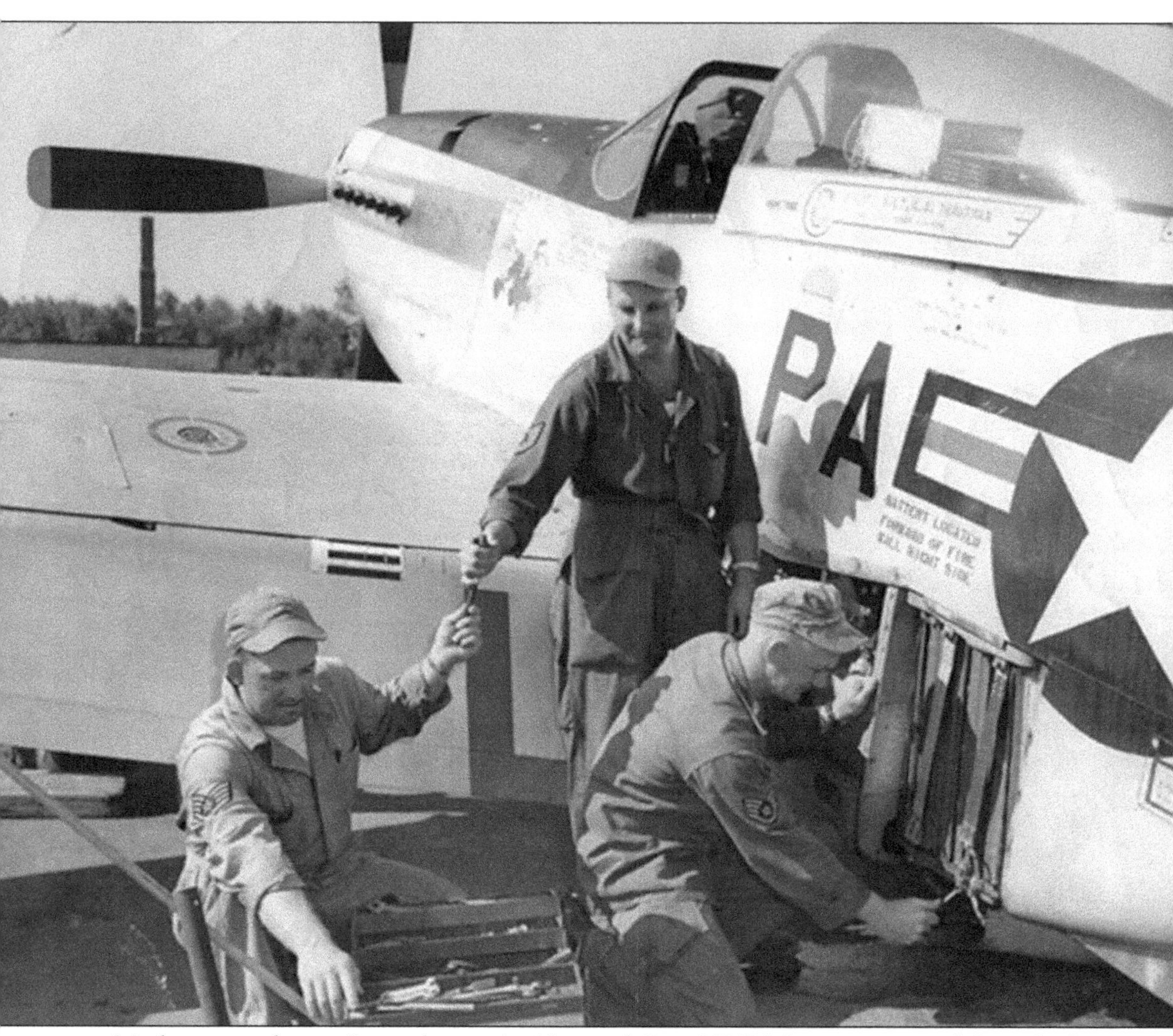

Mechanics work on a Pennsylvania Air National Guard F-51 Mustang, acquired from the Iowa Air National Guard, at Reading's 148th Fighter Squadron during the postwar era. The Reading squadron was reactivated on November 1, 1952, for the Korean War. Non–jet powered aircraft were still being used at the airport because the National Guard and Air Force runway length requirement to operate jets was 6,000 feet. Although public and military pressure was placed on the city to extend a runway, Reading's longest runway remained 5,150 feet in length, sealing the fate of a greater military presence in Reading. (Courtesy of Cyndi Thomas Deerfield.)

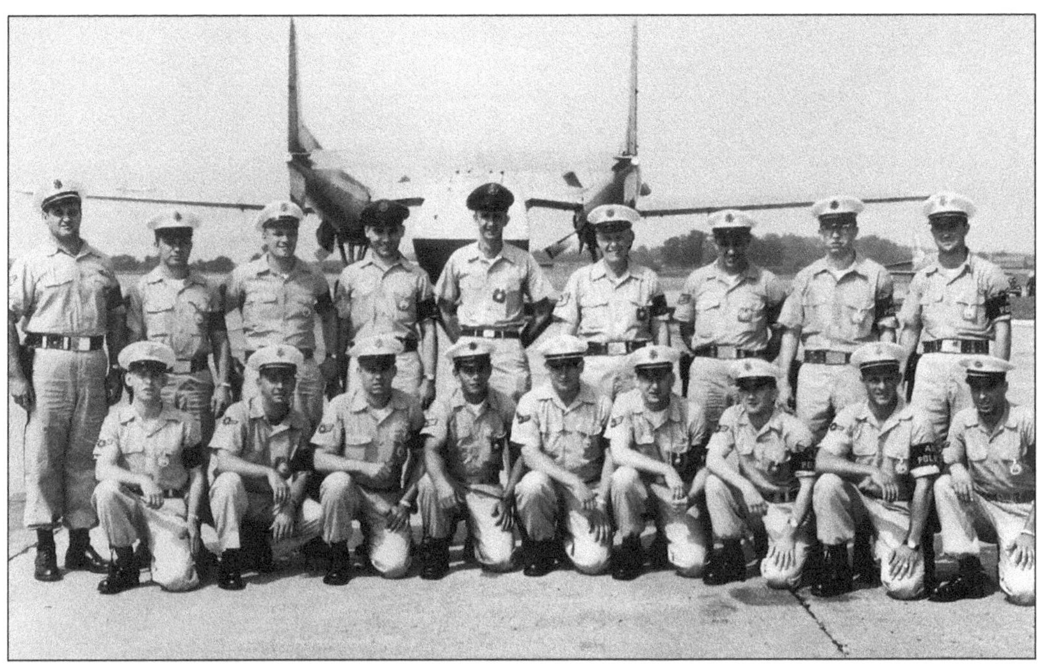

The Aeromedical Unit, Air Police Section of the 140th Pennsylvania Air National Guard is pictured above in front of a C-119 Flying Boxcar at Reading Airport in 1960. The unit was equipped with the Curtiss C-46, and in 1958, with the C-119. By 1960, the C-119 was replaced by the C-121 Constellation, but due to longer runway requirements, the Constellation was unable to fly out of Reading. The C-119, developed from the World War II Fairchild C-82, was designed to carry cargo, personnel, and patients, and drop cargo and troops by parachute. The first C-119 made its initial flight in November 1947, and by the time production ceased in 1955, more than 1,100 had been built. (Above, courtesy of Terry Sroka; below, courtesy of the US Air Force Museum.)

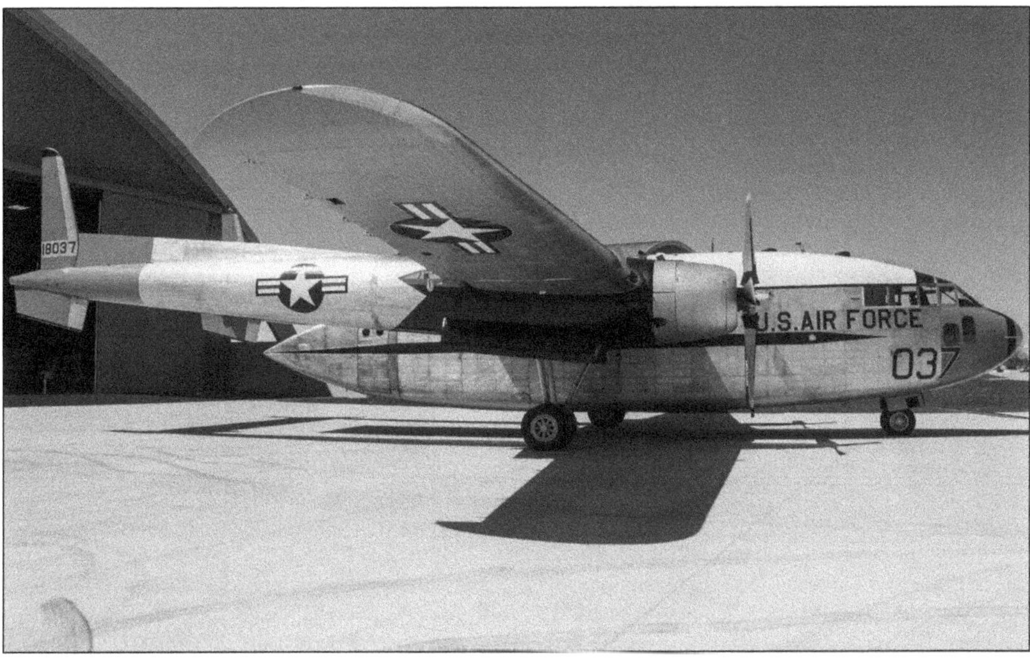

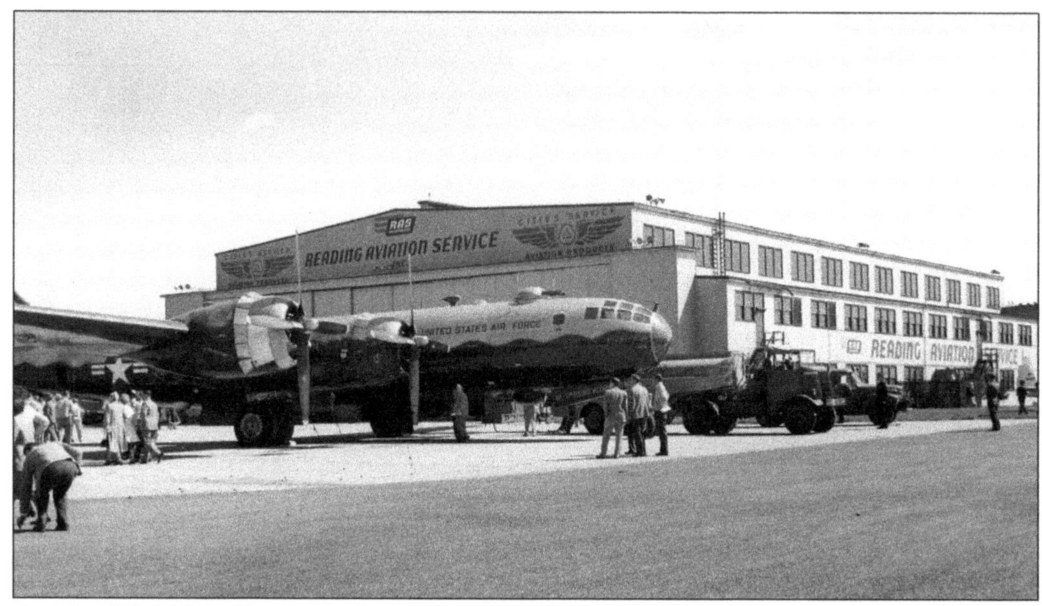

Another World War II aircraft, the Boeing B-29 Superfortress, sits next to the Reading Airport's Reading Aviation Service hangar in 1949. For such a large aircraft, many of its flying qualities were similar to smaller aircraft, which was important at a time when the airport's runway length restricted large aircraft operations. (Courtesy of Ron Ciervo.)

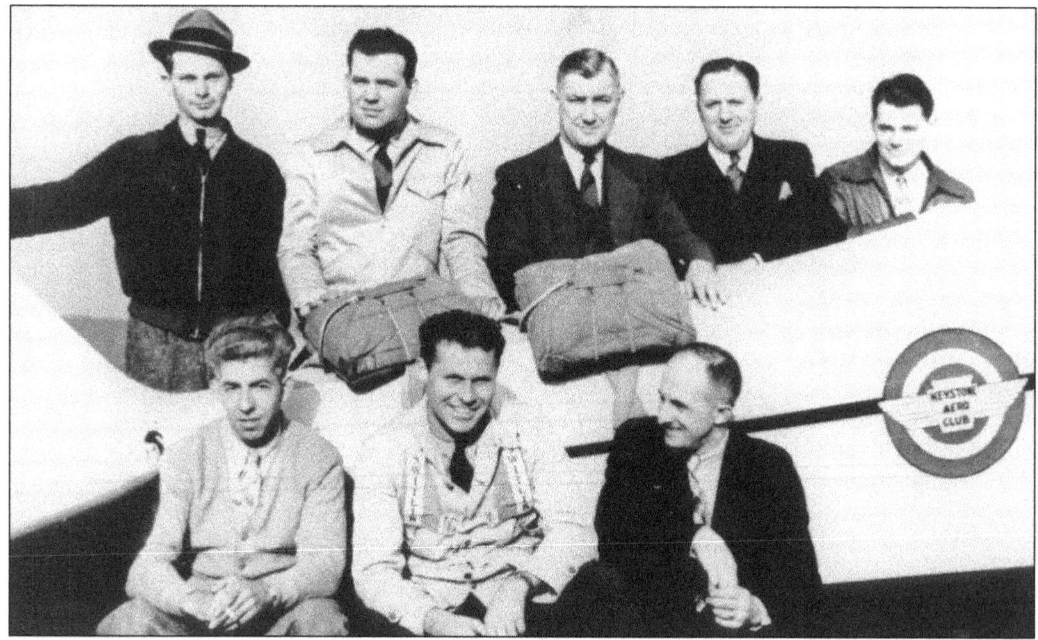

Although not considered military, the Pennsylvania State Police Flying Club operated the Keystone Aero Club's J-3 Cub out of Madeira Field. Pictured from left to right are (first row) Fritz "The Fox" Mengle, Sime Bertolet, and Sergeant Muck; (second row) Officer Hall, Sergeant Sickle, and three unidentified. (Courtesy of Sime Bertolet.)

Seven
COMMERCIAL AND GENERAL AVIATION

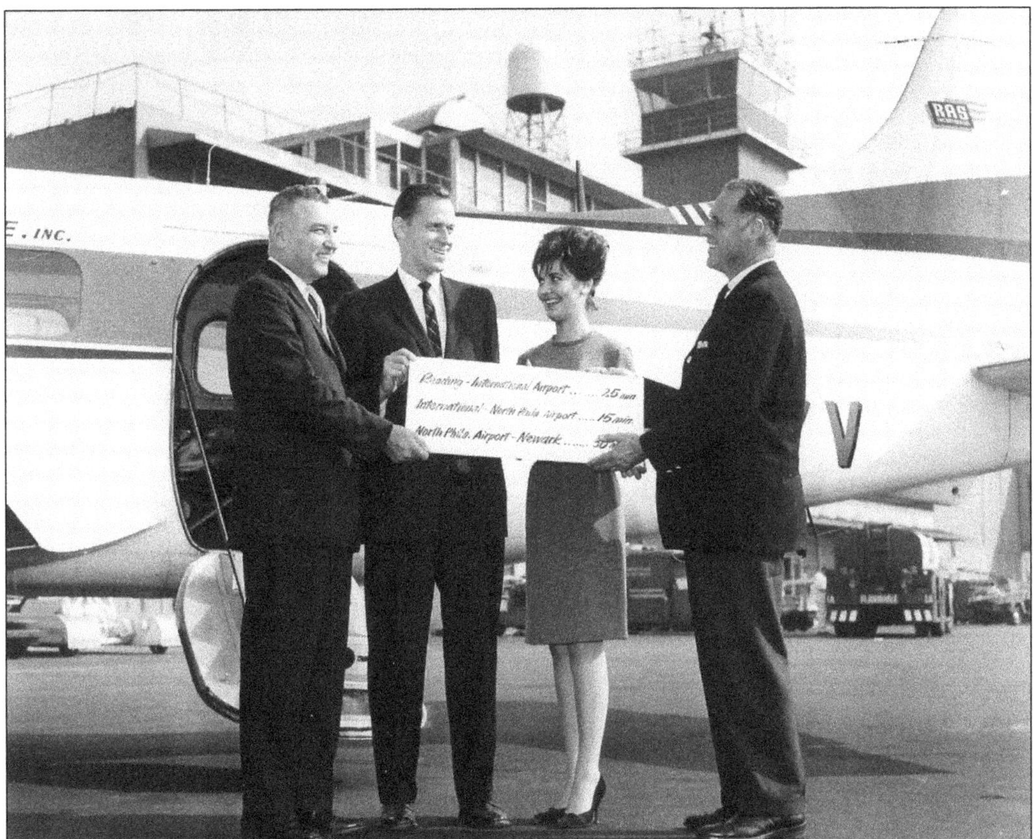

Reading's aviation prominence began in the 1920s with Reading Airways as the first commercial flight operation in Pennsylvania. Since then, Reading Air Service became a leading aviation operation led by Alfred M. "Sime" Bertolet, R. Harding Breithaupt, and Brooks McElroy after World War II with one division operating as a charter airline. In 1969, the charter operation was renamed Suburban Airlines; by 1973, Suburban became a member of Allegheny Commuter. (Courtesy of Sime Bertolet.)

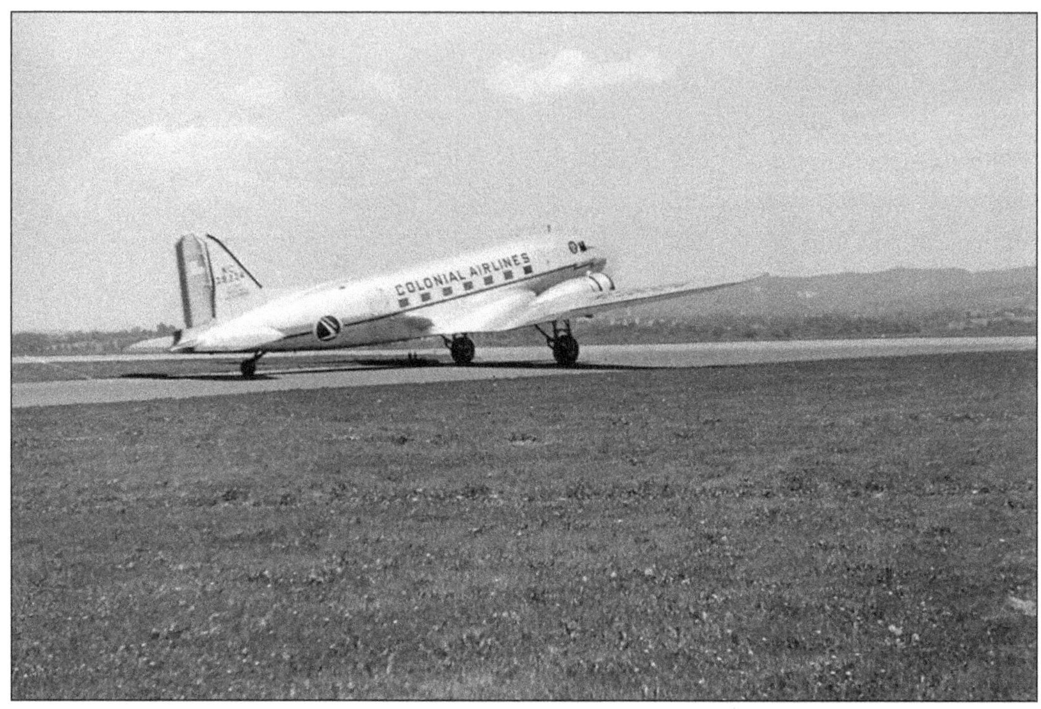

Founded as Canadian Colonial Airways in 1928, by 1942, the airline was renamed Colonial Airlines. It would eventually merge with another airline in 1956 to become Eastern Airlines, with flights to airports such as New York, Baltimore, and other regionalized destinations. These 1940s images shows a Colonial DC-3 taxiing at Reading Airport. (Both, courtesy of Terry Sroka.)

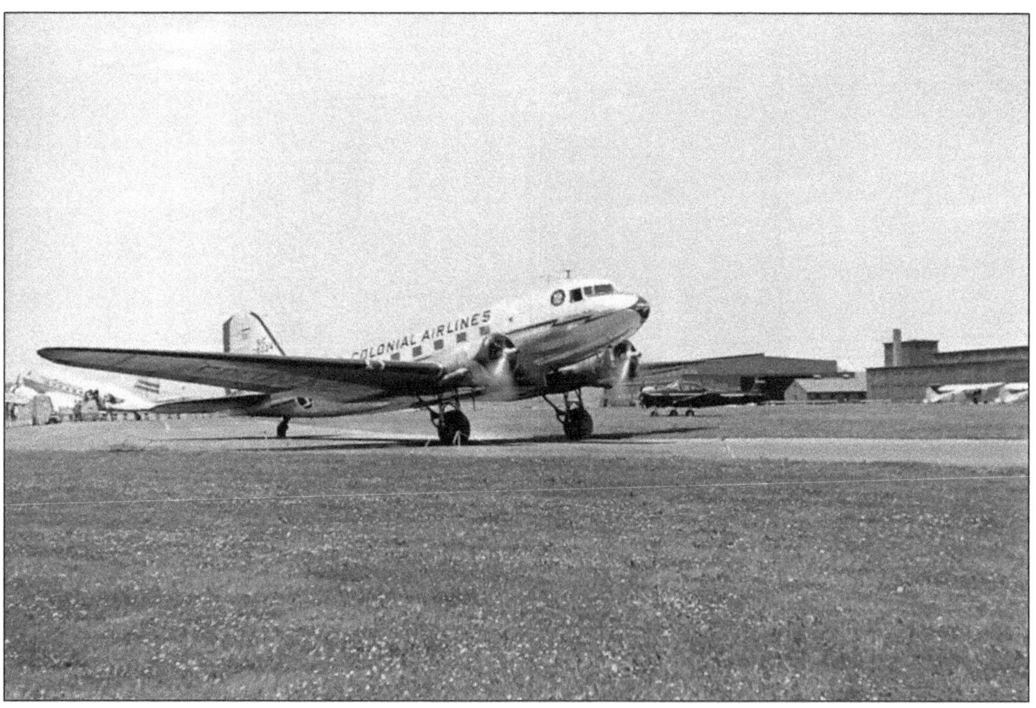

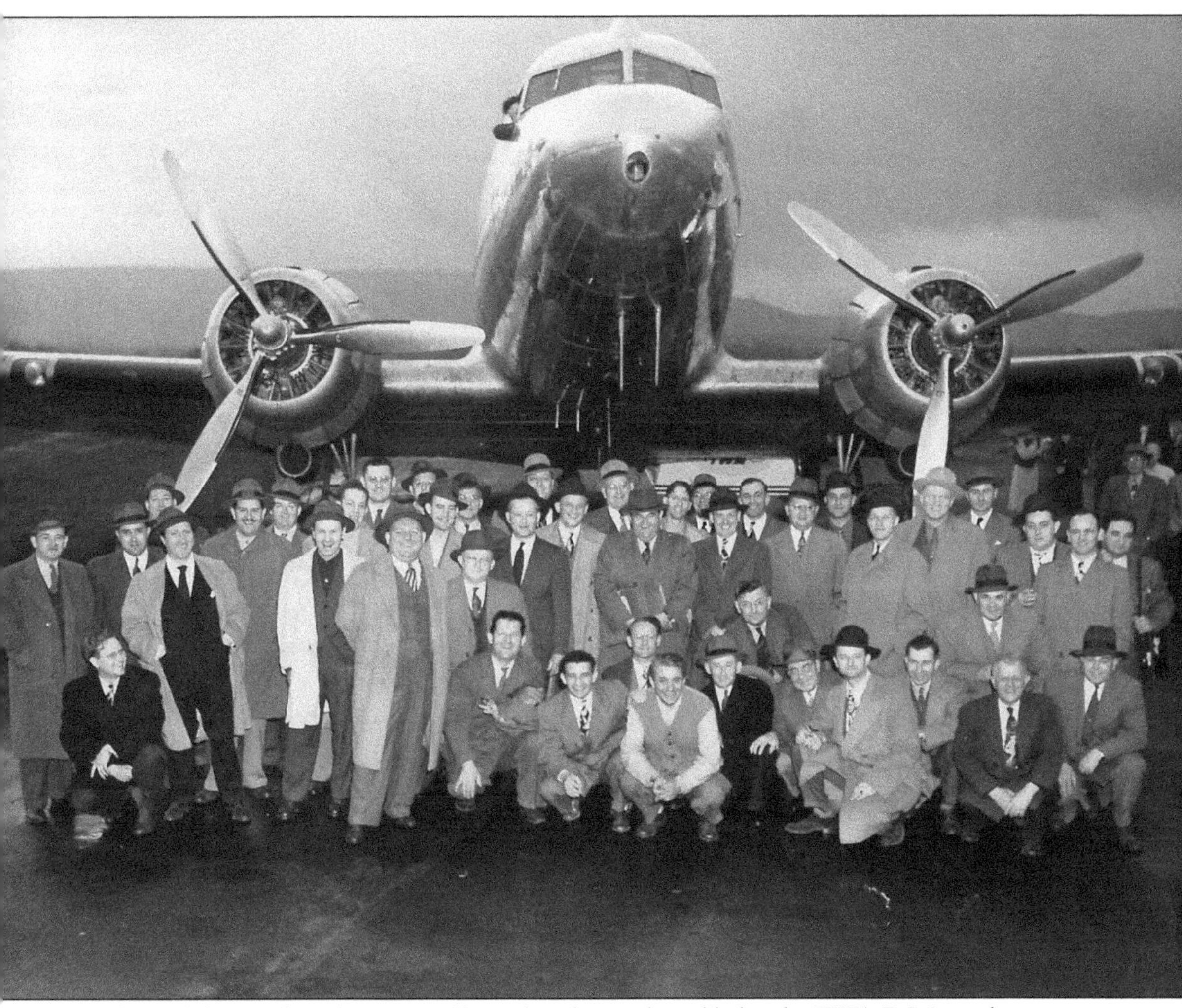

Members of the Reading Airport staff stand underneath and behind a TWA DC-3 on the main ramp during the 1940s. Douglas built 455 commercial DC-3 airplanes, and 10,174 were produced for the military as the C-47, serving as a transport during World War II. (Courtesy of the Berks History Center.)

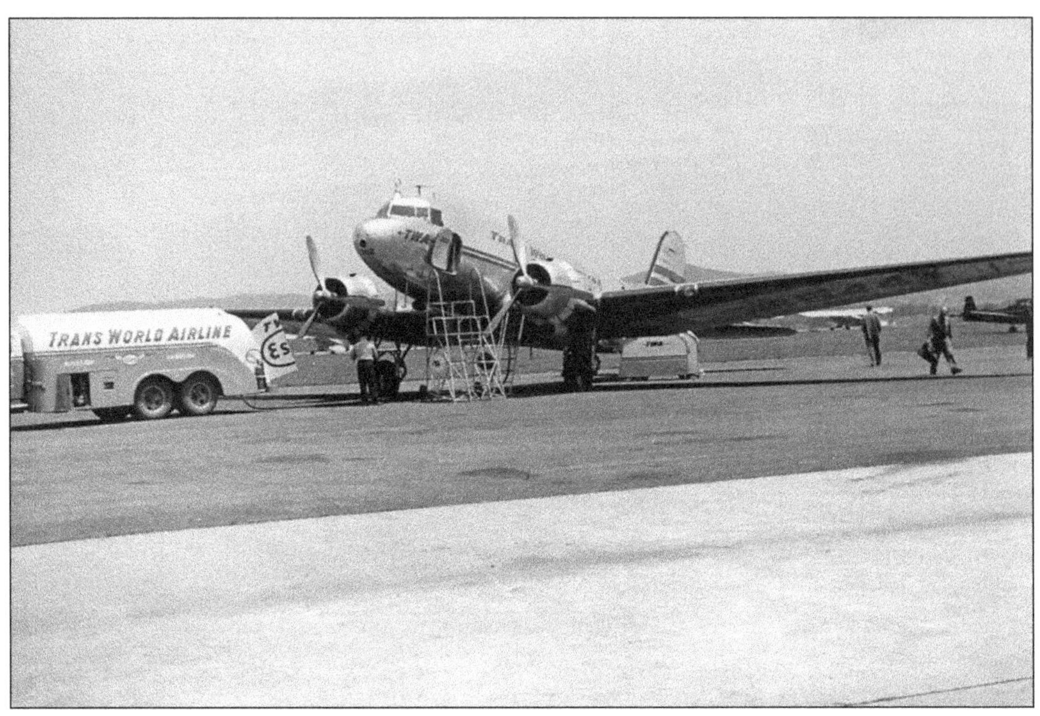

TWA was another airline that operated out of Reading Airport, with service beginning on August 1, 1941, connecting Reading with other cities on the east-west route from coast to coast. Although the DC-3 was a popular airplane during its time, Reading also had larger TWA aircraft, such as the Lockheed Constellation, flying from the field. (Both, courtesy of Terry Sroka.)

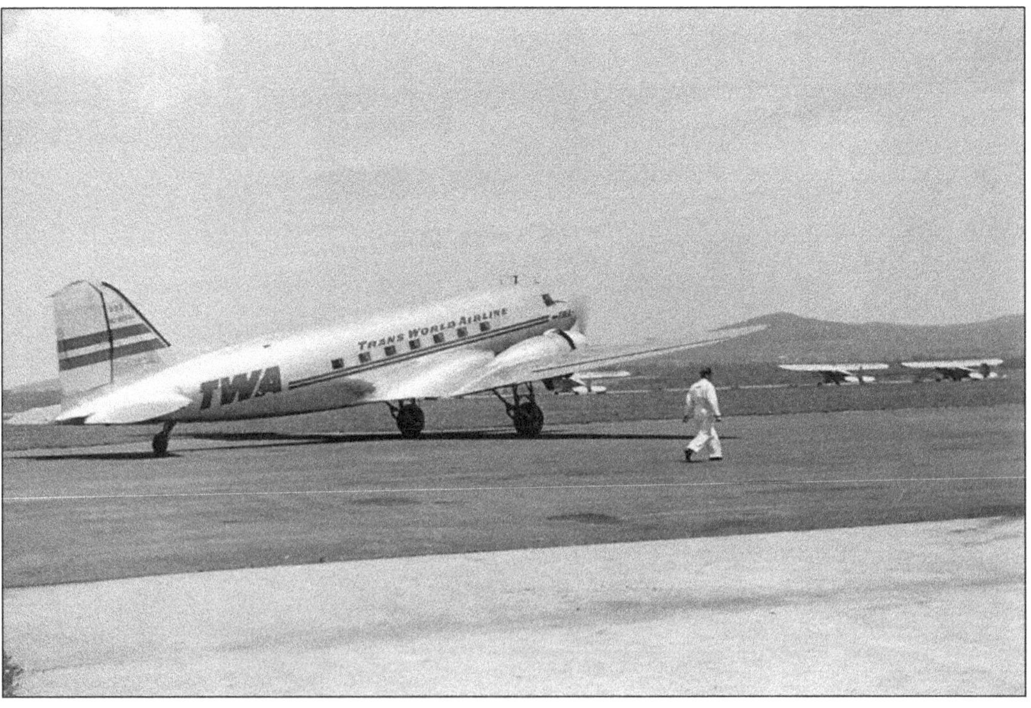

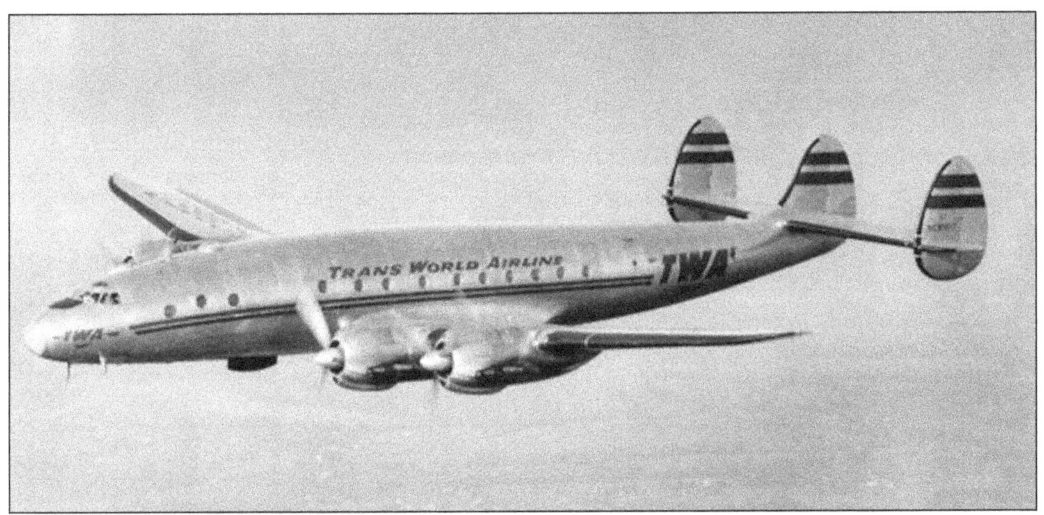

The first crash of a Lockheed Constellation, the world's largest passenger plane during the mid-1940s, occurred on July 11, 1946, when a TWA L-049 Constellation named *Star of Lisbon* departed from the Reading Municipal Airport and climbed to 3,000 feet four miles east of the airport to conduct practice instrument approaches. The aircraft was based at the Reading Airport for training transoceanic crew members. The training program began in January 1946, and NC86513 was one of four Constellations assigned to the program. Shortly after takeoff, the flight crew detected an odor resembling burning insulation. The flight engineer went aft to determine the origin of the smoke, and upon opening the galley door, smoke filled the cockpit, resulting in the accident. (Above, courtesy of Terry Sroka; below, courtesy of AP Images.)

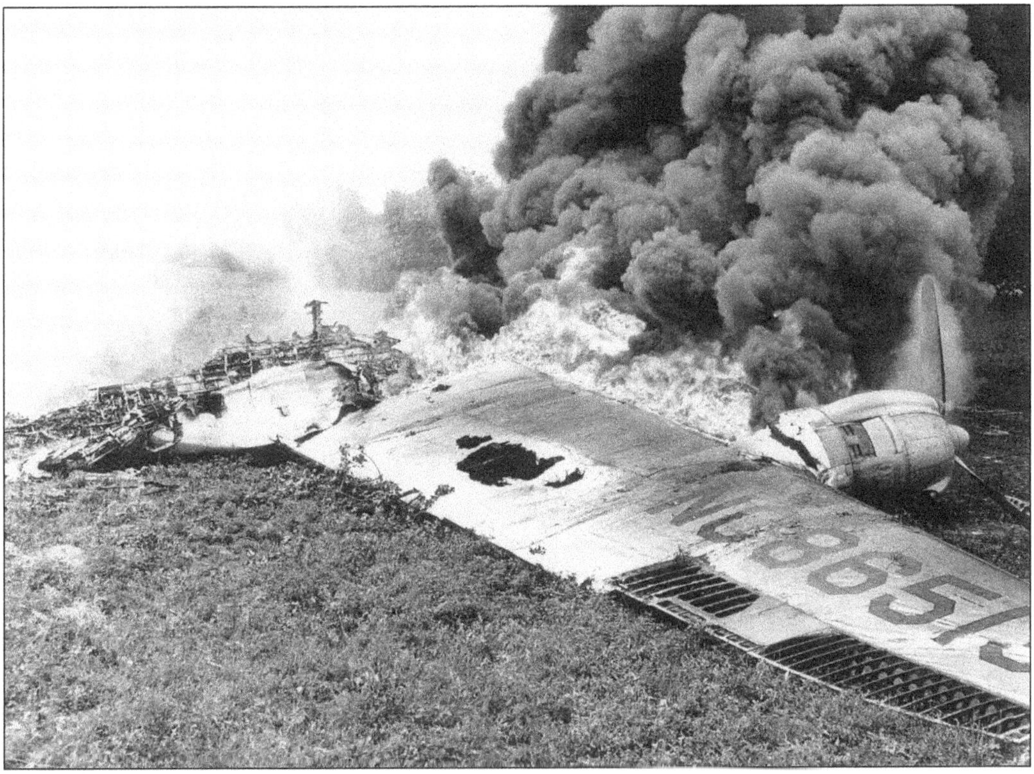

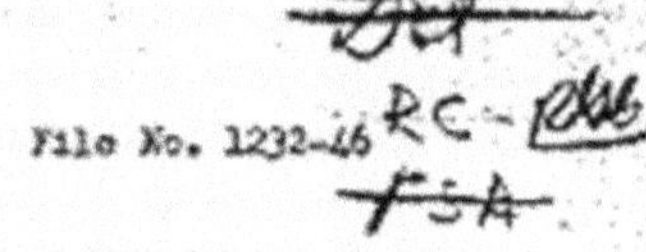

Docket No. SA-120 File No. 1232-46

CIVIL AERONAUTICS BOARD

ACCIDENT INVESTIGATION REPORT

Adopted: November 29, 1946 Released: December 18, 1946

TRANSCONTINENTAL AND WESTERN AIR, INC., READING, PA., JULY 11, 1946

The Accident

NC 86513, owned and operated by Transcontinental and Western Air, Inc., crashed approximately one mile northeast of Reading Airport, Reading, Pennsylvania, at 1140, Eastern Standard Time*, July 11, 1946, as a result of fire in the forward baggage compartment which produced smoke of such density that the pilot was unable to maintain sufficient control of the aircraft to effect a normal landing. Five crew members were fatally injured; the instructor pilot was seriously injured; and the Lockheed Model 049, Constellation, was demolished by impact and fire**.

History of the Flight

NC 86513, hereinafter referred to as Aircraft 513, was assigned to Reading Airport for use in Constellation transition in the training program of the International Division of TWA. Aircraft 513, with

---

* All time referred to herein is Eastern Standard and based on the 24-hour clock.

**The following report is a summary of the findings of the investigation of this accident and the detailed examination of the aircraft involved. Since the date of the accident, however, extensive modifications of the Lockheed Model 049 have been accomplished in order to eliminate the mechanical deficiencies which contributed to this accident. The primary cause of this accident has been eliminated. Other deficiencies revealed by the accident have been eliminated or are under further study. A resume of the modifications which have been accomplished is appended hereto. (See page 28.)

According to the official accident report, the probable cause was electrical arcing from a generator into the fuselage skin of the forward baggage compartment, which resulted in intense local heating. The burning of the fuselage insulation produced smoke of such density that control of the aircraft was impossible. Following the accident, the Civil Aeronautics Administration ordered that all Lockheed Constellations be grounded for 30 days. This was the first time a major airliner was grounded by government order. The accident was featured in the movie *The Aviator* about Howard Hughes, who oversaw operations at TWA and was the designer of the Constellation in 1939. The order grounded overseas and domestic flights, causing significant operational losses to the airlines flying the Constellation. (Courtesy of the National Air and Space Museum.)

Capital Airlines served the eastern United States before merging into United Airlines in 1961. At its peak, it was the fifth largest domestic carrier in the United States. In 1950, Capital received its first Lockheed Constellation, enabling it to compete more effectively on longer distance routes. By 1955, it became the first US operator of Vickers Viscounts—the first passenger turboprop aircraft. (Right, author's collection; below, courtesy of National Air and Space Museum.)

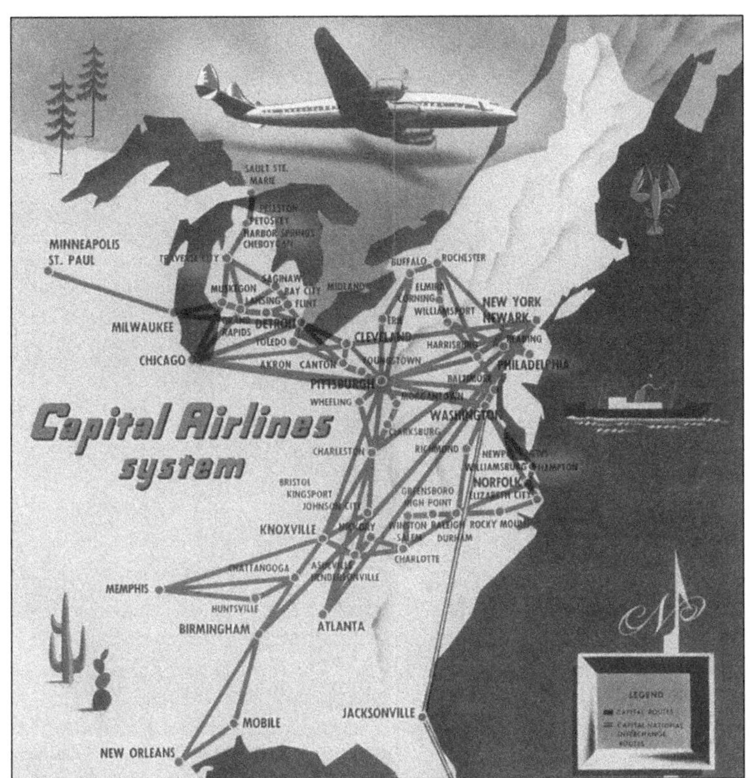

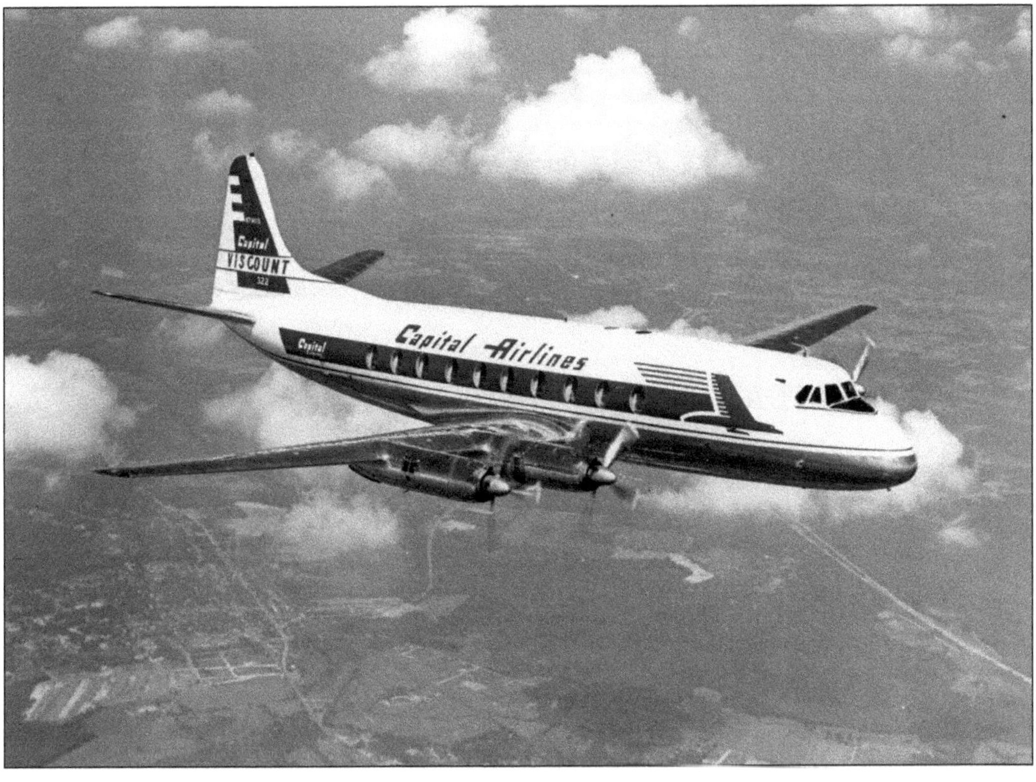

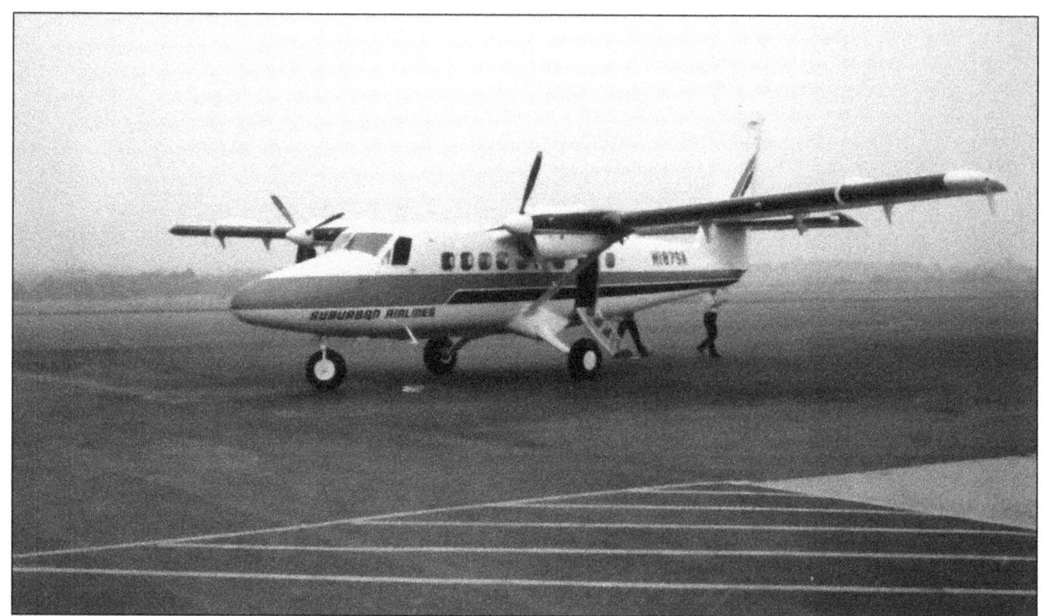

Suburban Airlines was one of many commuters with a substantial history prior to deregulation. It could trace its history to Reading Aviation Service in 1957. It became Suburban in 1968 and joined the Allegheny Commuter network in 1973. By 1980, all services operated under the Allegheny Commuter brand were acquired by USAir. Pictured in the 1960s, this De Havilland Canada DHC-6 Twin Otter 200, registration N187SA, was sold to another carrier and crashed at sea, killing all aboard. (Courtesy of Terry Sroka.)

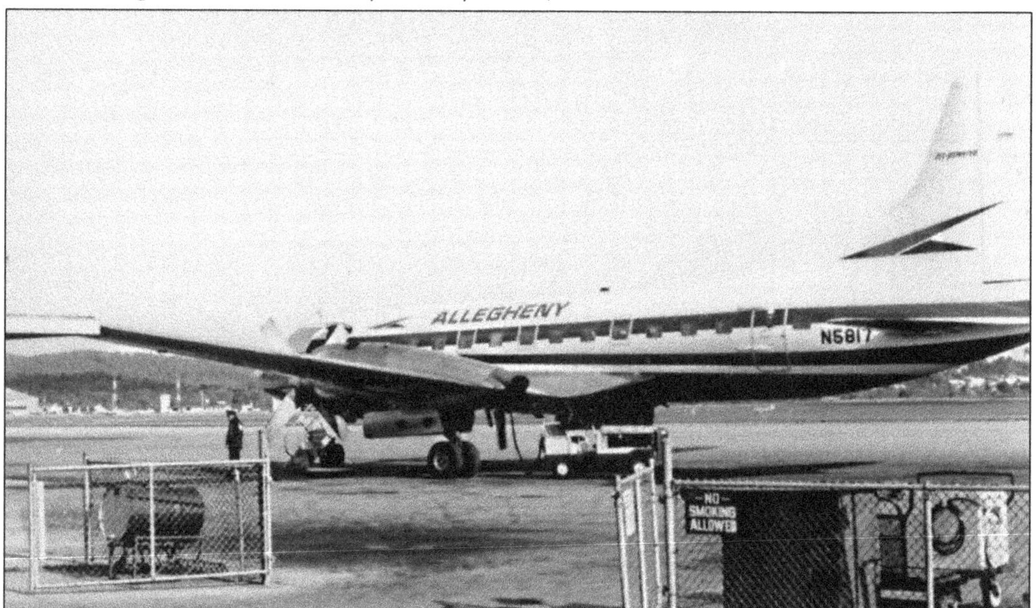

Allegheny Airlines was one of 10 regional airlines operating as US Airways Express from Reading Airport. Allegheny was made up of two smaller operators: Pennsylvania Commuter Airlines in Harrisburg and Allegheny Commuter Airlines headquartered at the Reading Airport. This Convair 580, seen in front of the Reading Airport terminal ramp in 1966, was the first conversion from piston to turboprop. This model first entered service on January 19, 1960. (Courtesy of Terry Sroka.)

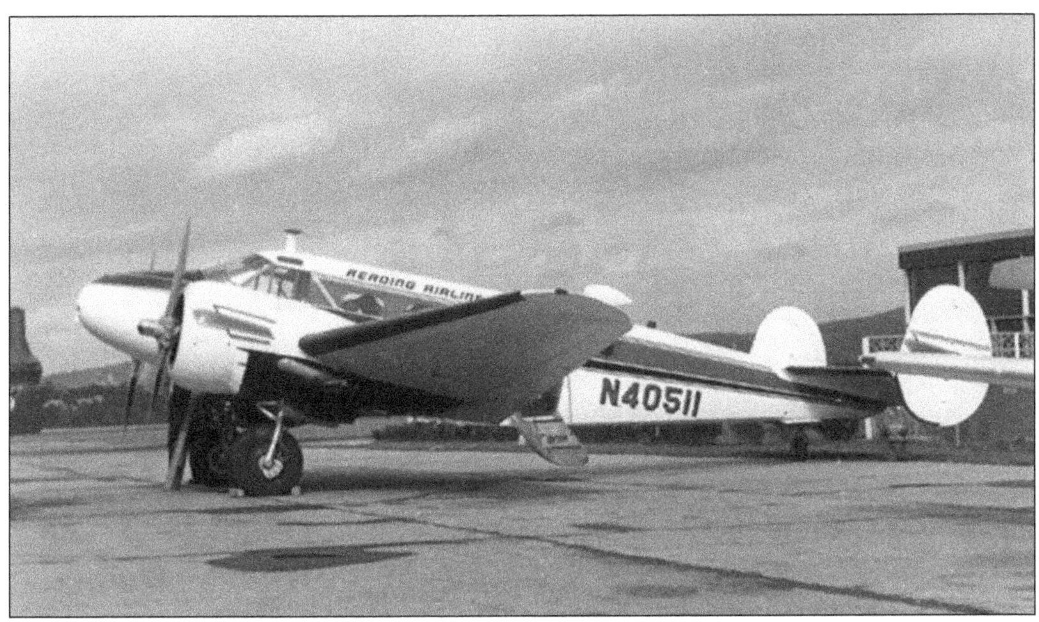

Reading Airlines operated rare aircraft, such as one of two De Havilland DH104 Dove 6BAs, in its fleet, as seen above in 1960 at the Reading Municipal Airport. The aircraft was a British short-haul airliner introduced in 1946. The Reading Airlines 1949 Beech D18S seen below in 1960 flew commuter service for up to 10 passengers from Reading to Newark, New Jersey, from 1958 to 1968. Reading Aviation Service was a full-service fixed base operator serving Reading and surrounding airports. (Both, courtesy of Terry Sroka.)

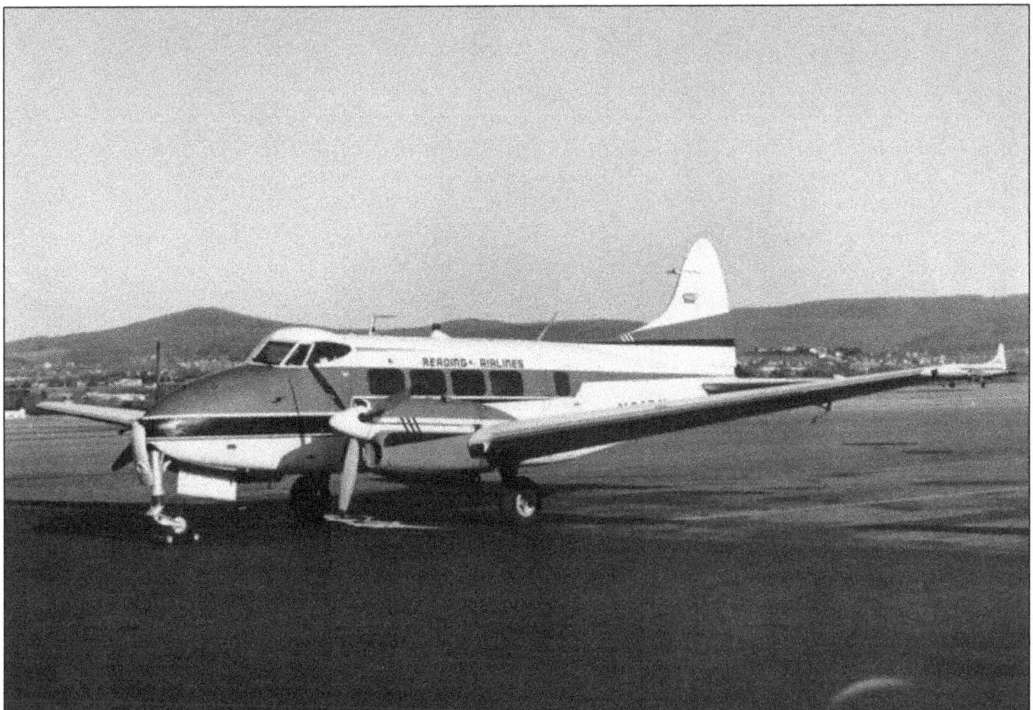

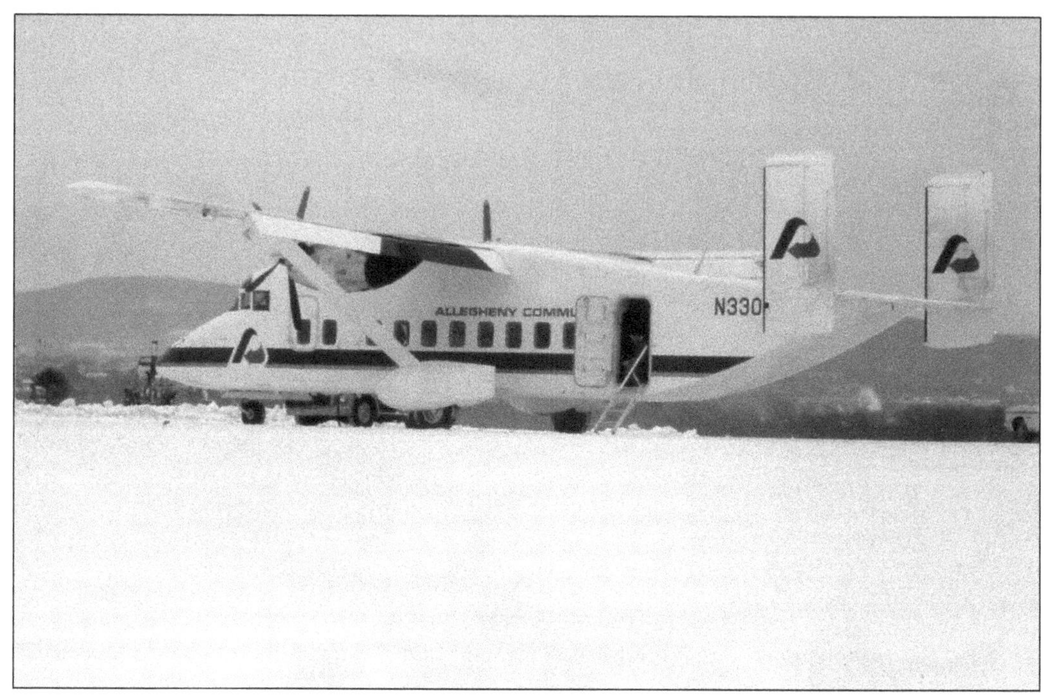

Allegheny Airlines operated the Short 330, shown above on a snowy ramp, as part of its regional turboprop aircraft fleet. The Short SC7 Skyvan Series 2, shown below with children sitting on the grass at the 1967 Reading Airshow, was a shorted version of the Short 330 and 360 used by Allegheny during the late 1970s and early-to-mid-1980s. (Above, courtesy of Terry Sroka; below, courtesy of the National Air and Space Museum.)

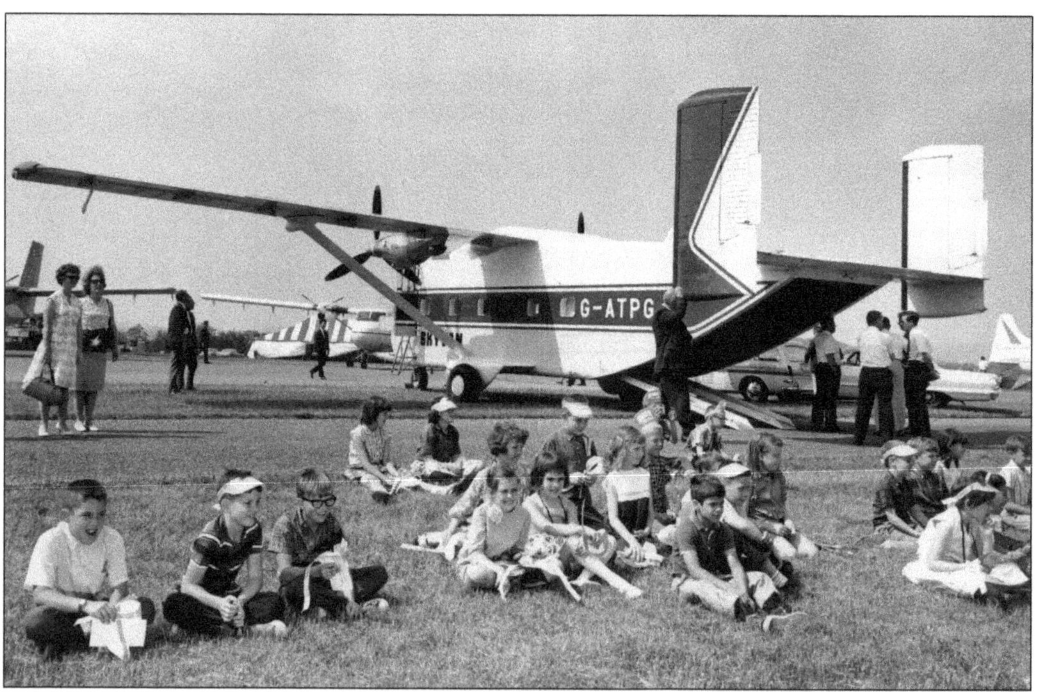

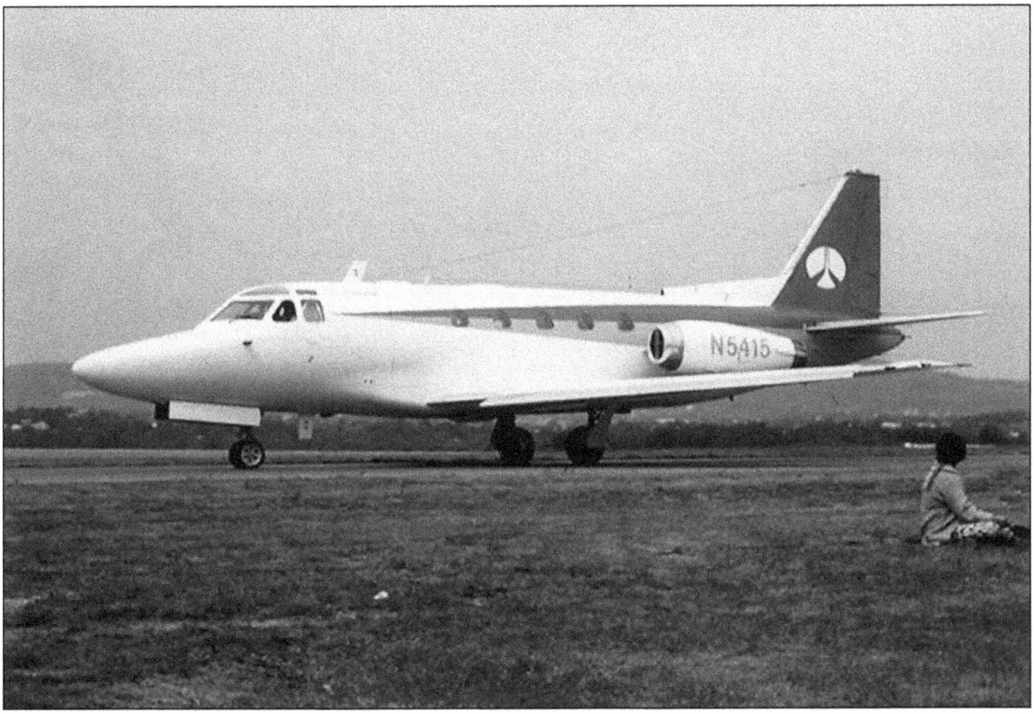

The Rockwell Sabreliner 60A seen above at Reading Airport in 1969 was a popular mid-sized business jet and was also used by the military. Below is one of the first Lockheed JetStars on display at the 1963 Reading Airshow. The JetStar was the first business jets to enter service and was a leader in its class for many years. (Both, courtesy of Larry Green.)

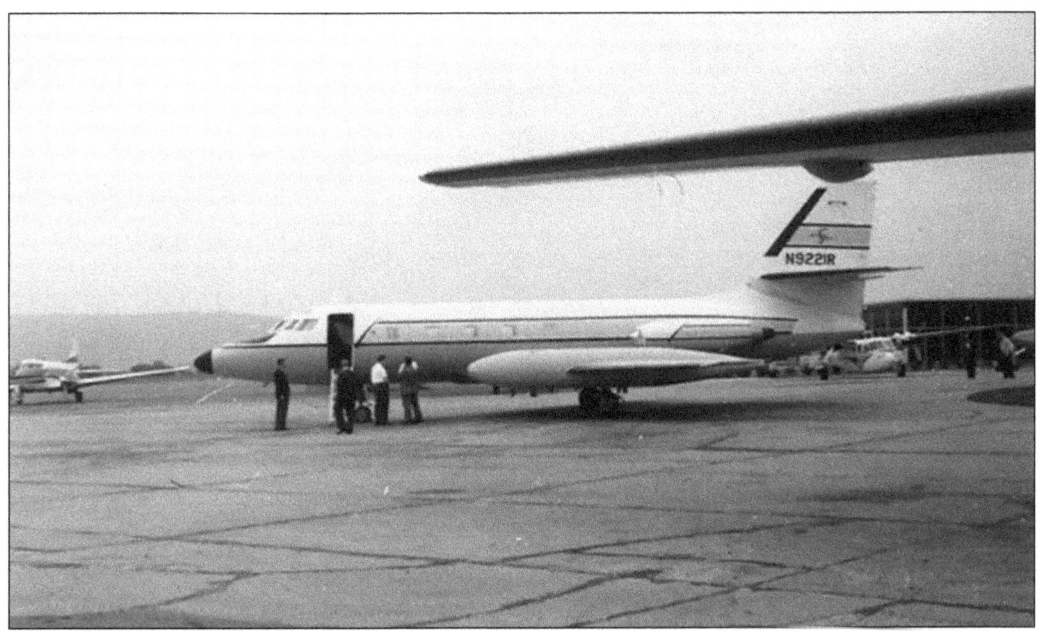

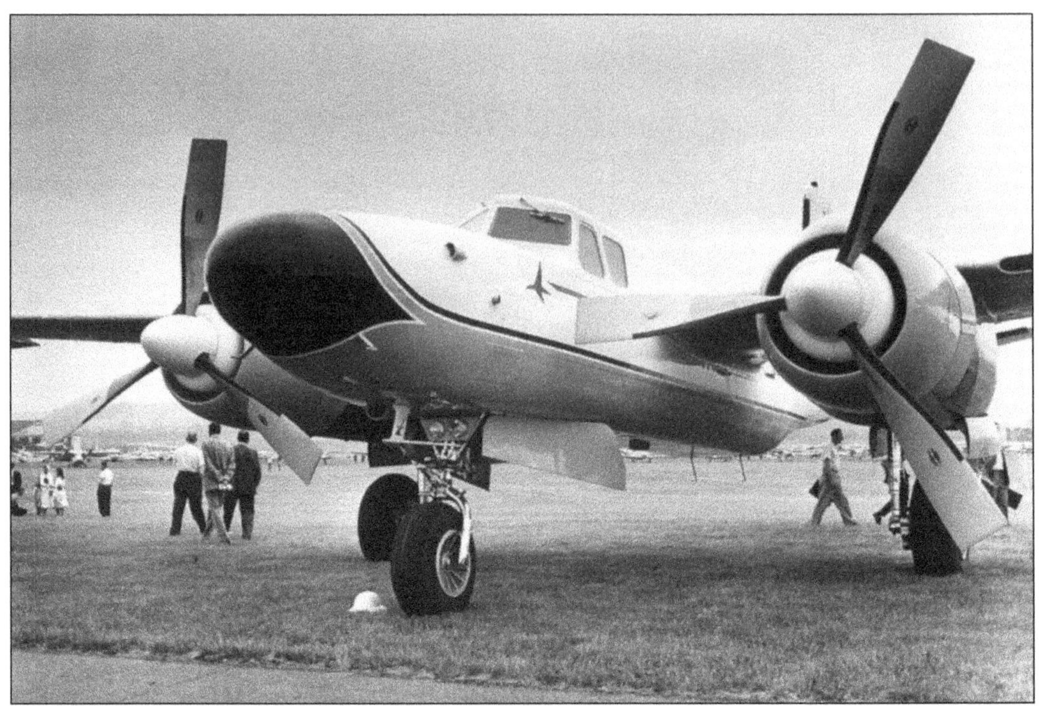

The Douglas A-26 Invader seen above at the Reading Airshow in 1961 was owned by Weaton Glass of Millville, New Jersey. The Invader was used by the military between 1948 and 1965 as a light bomber and ground attack aircraft. Below, outside the Reading Aviation Service hangar is an Aero Commander 680E. RAS was a Rockwell Aero Commander dealer and most likely used this 1974 photograph for marketing purposes. (Both, courtesy of Larry Green.)

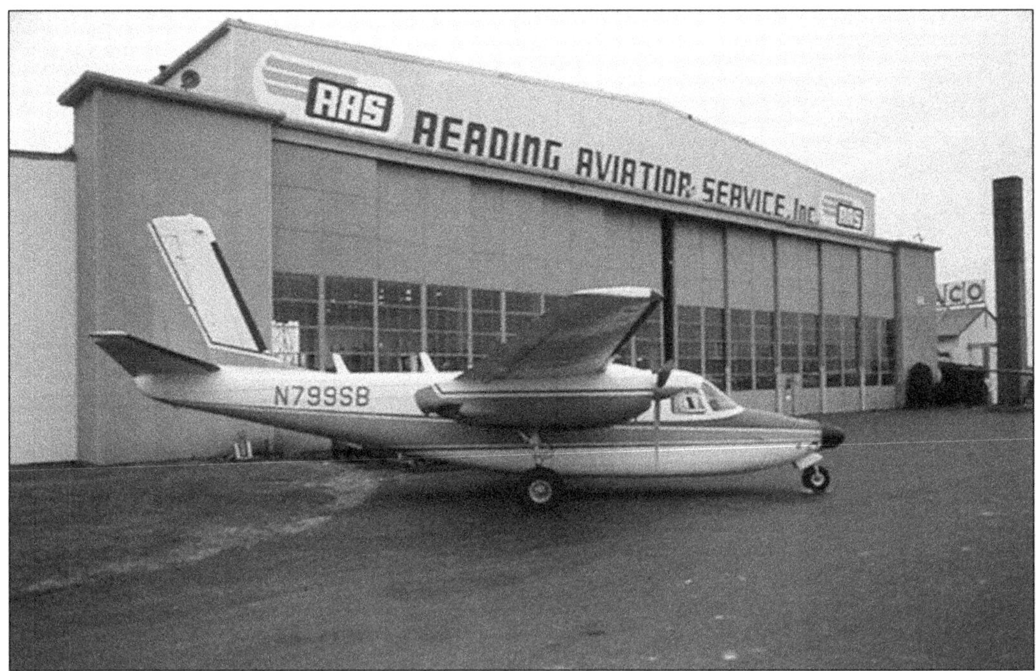

Eight

Aviation-Related Industries

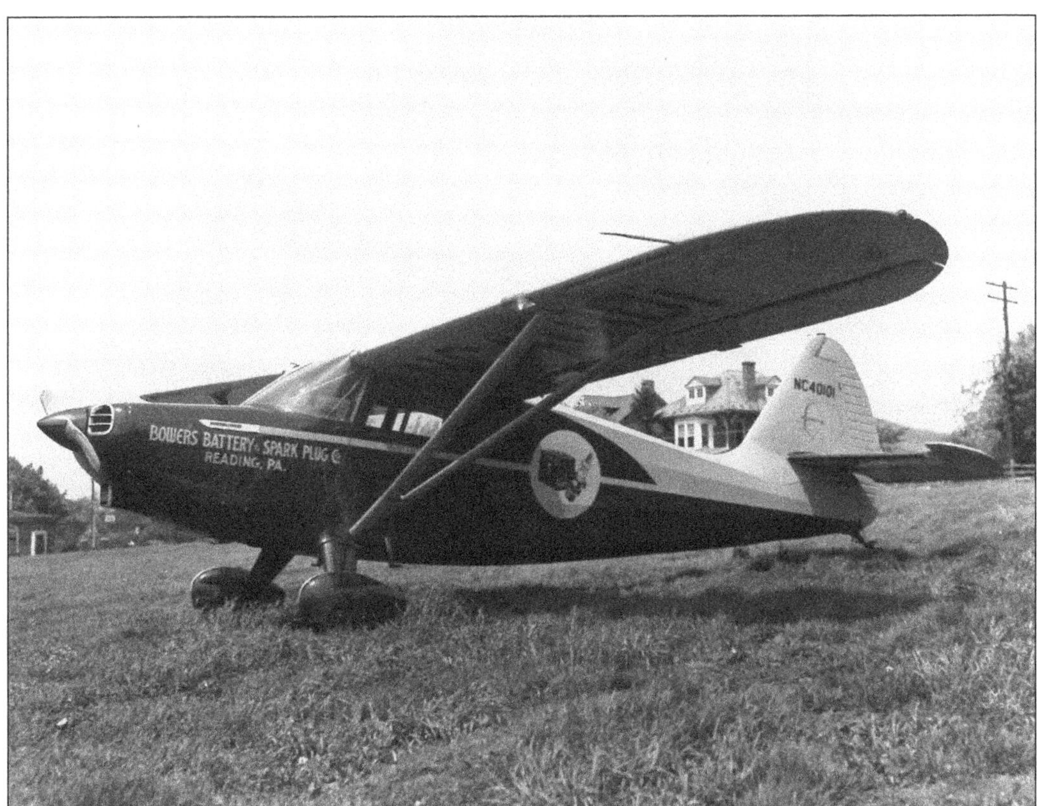

Berks County's industries played an important role in many areas of aviation. These included steel for building aircraft structures, armaments, aircraft batteries, textiles, engine parts, and other aviation-related products that helped the United States military. Because of the county's contribution to war efforts, the area experienced heightened security to ensure its protection. (Courtesy of Berks History Center.)

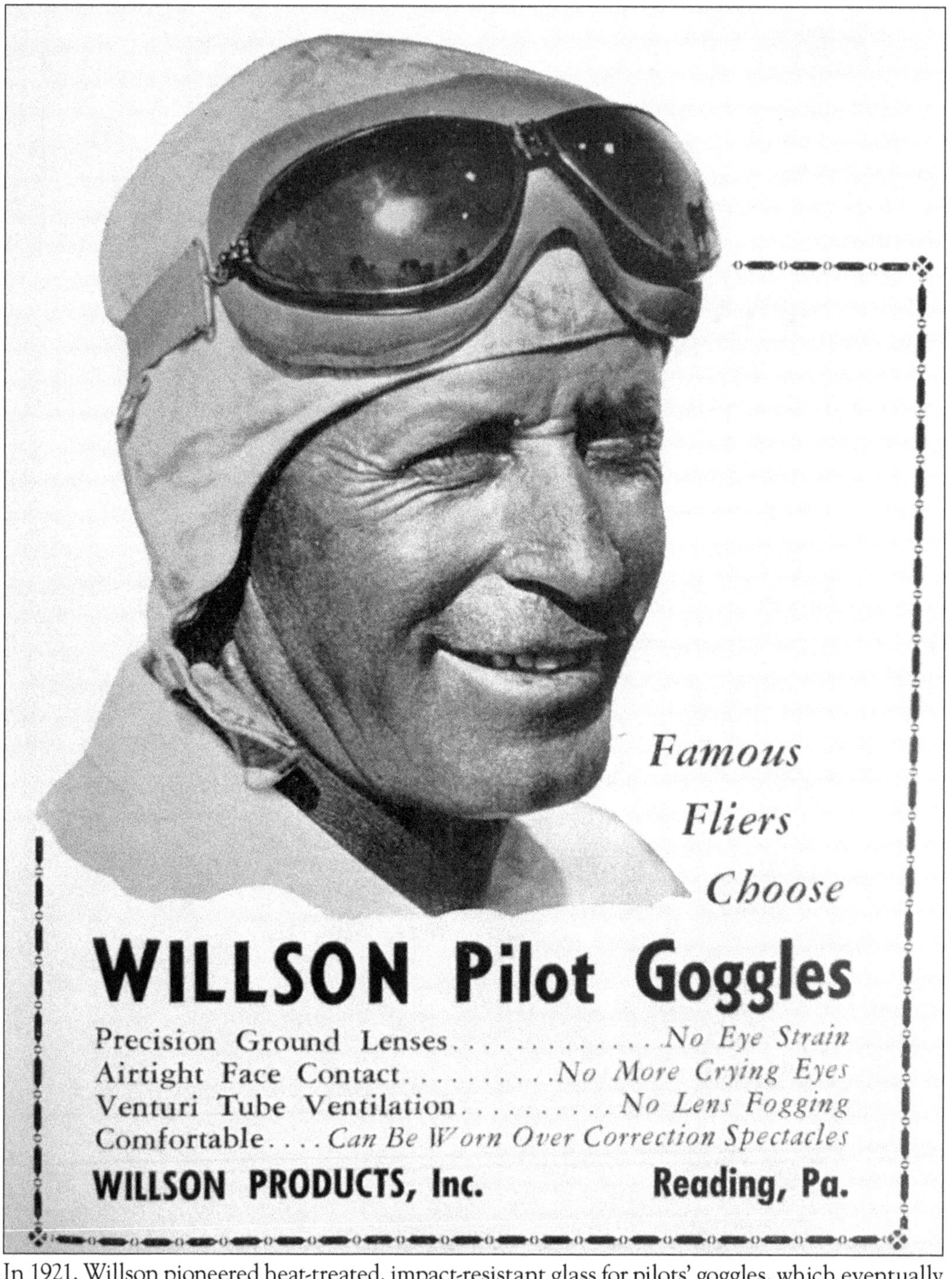

In 1921, Willson pioneered heat-treated, impact-resistant glass for pilots' goggles, which eventually helped the war effort by making aviator goggles and high-altitude oxygen masks for pilots in the military. By 1943, the factory manufactured all of the sunglasses issued to the US military during World War II. The company reached its peak during this time, employing 1,300 workers. (Courtesy of the Reading Airport.)

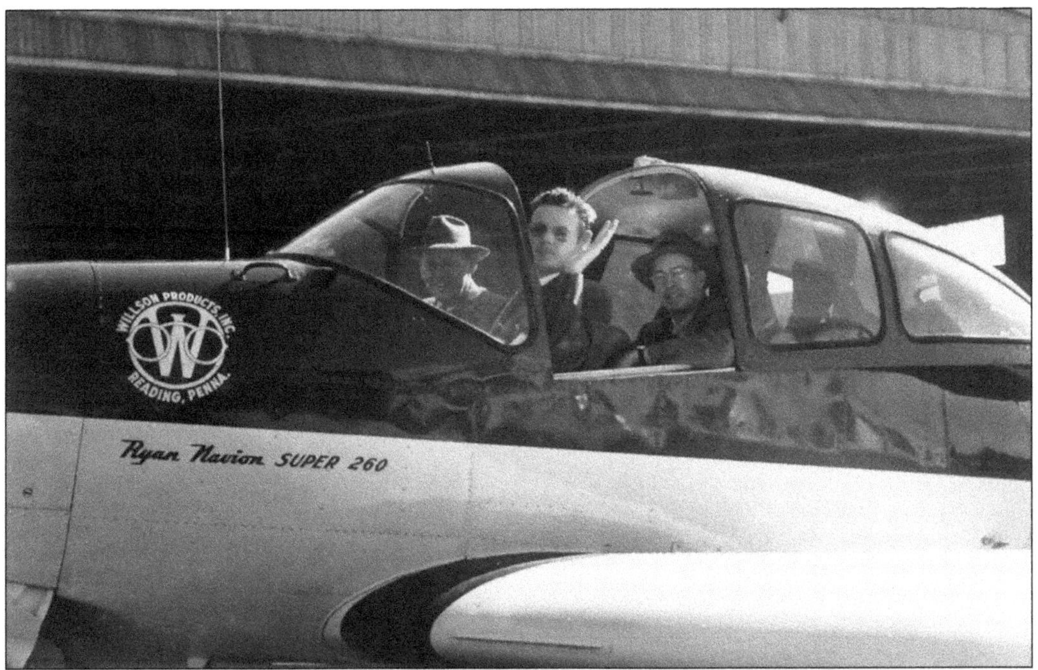

The photograph above shows Willson Products' Ryan Navion Super 260 at the Reading Airport, while below is a press photograph taken in the 1950s promoting Willsonite sunglasses. The Willsonite sunglasses were used by many airlines, the Army, and the Navy. Additionally, they were one of the most popular brands of personal sunglasses during the 1940s and 1950s. (Both, courtesy of Terry Sroka.)

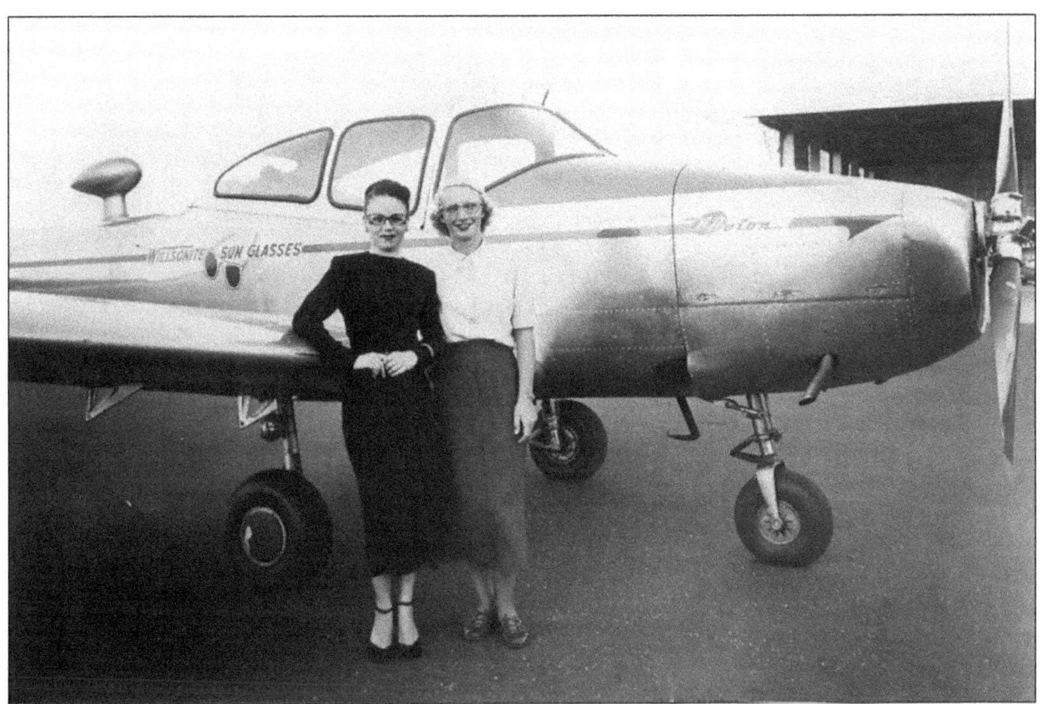

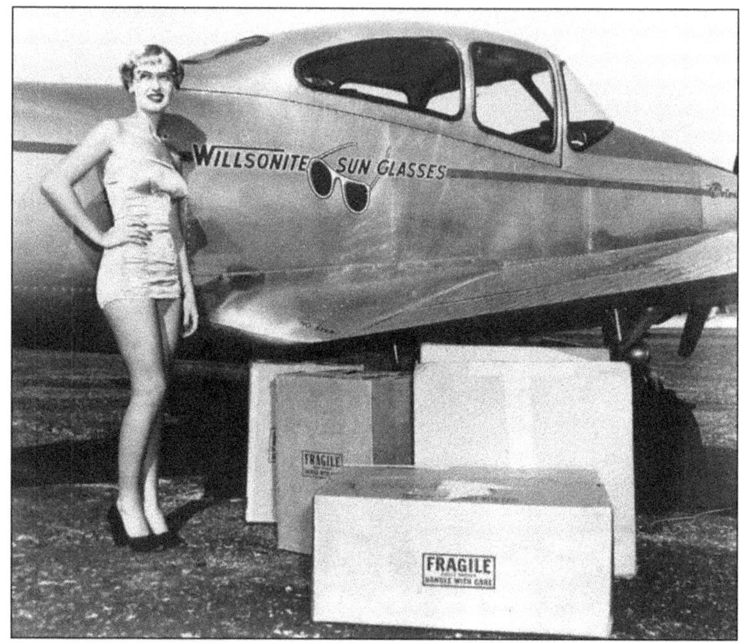

Here are more promotional photographs of the Willson Navion promoting safety glasses and Willsonite sunglasses manufactured in Reading. Today, Goggleworks occupies the site of the first factory in the world to manufacture optical glass for lenses and reading glasses. (Both, courtesy of Terry Sroka.)

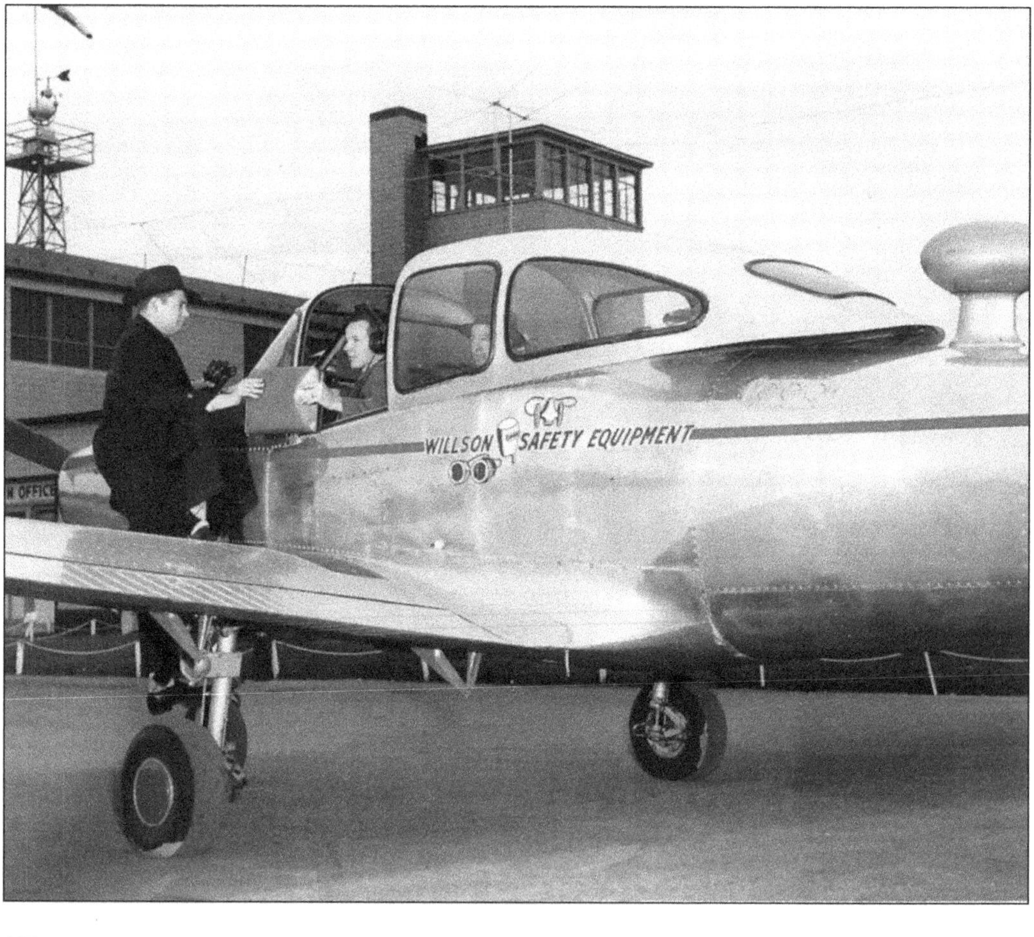

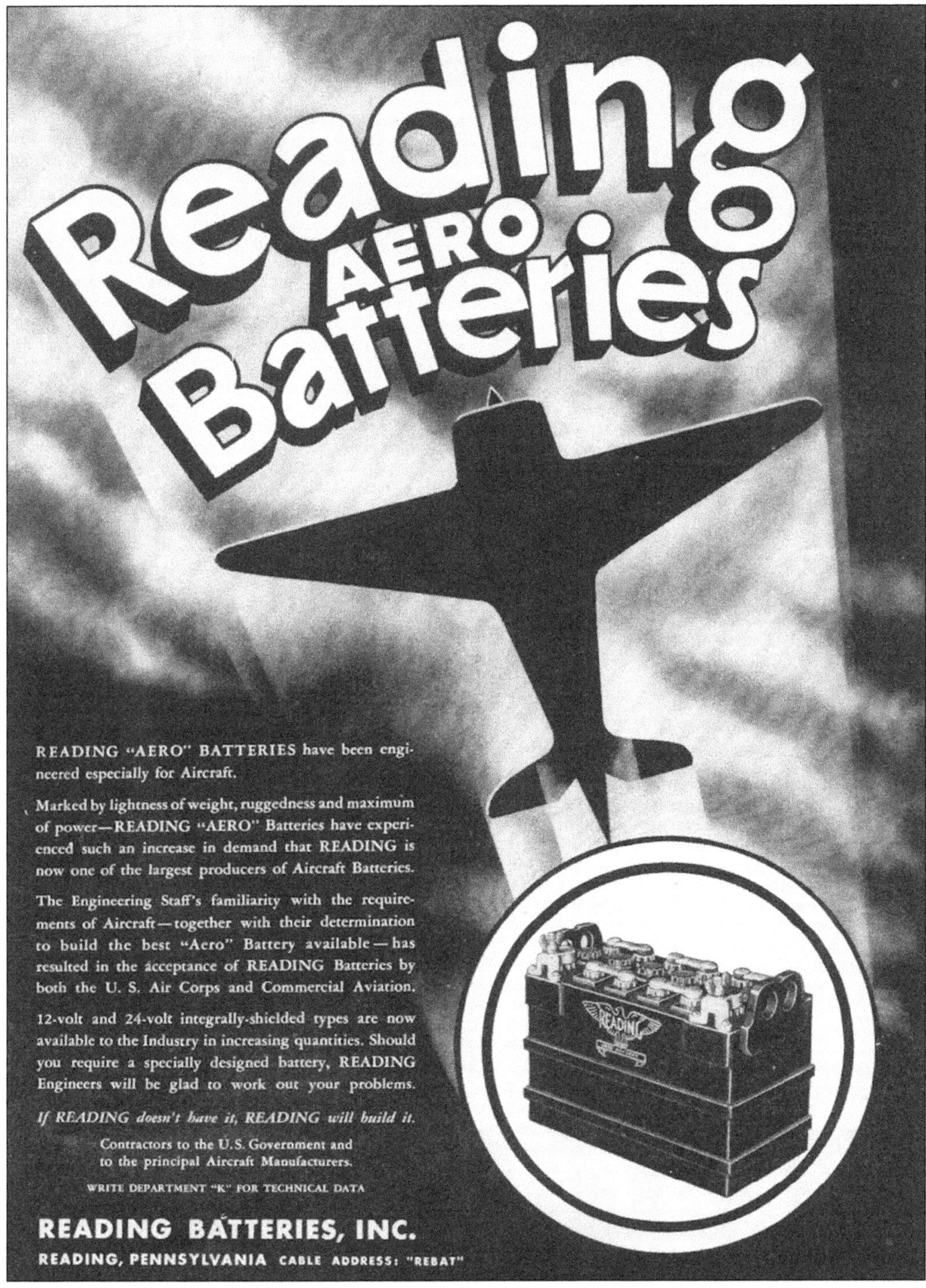

Reading has always been known as a location for manufacturing batteries. Reading Batteries Inc., or Rebat, as it was known, provided batteries for many applications, including aircraft. Rebat sponsored various events at the Reading Airport, which included the popular Rebat airplane races. (Courtesy of Gerald VanWyngaarden.)

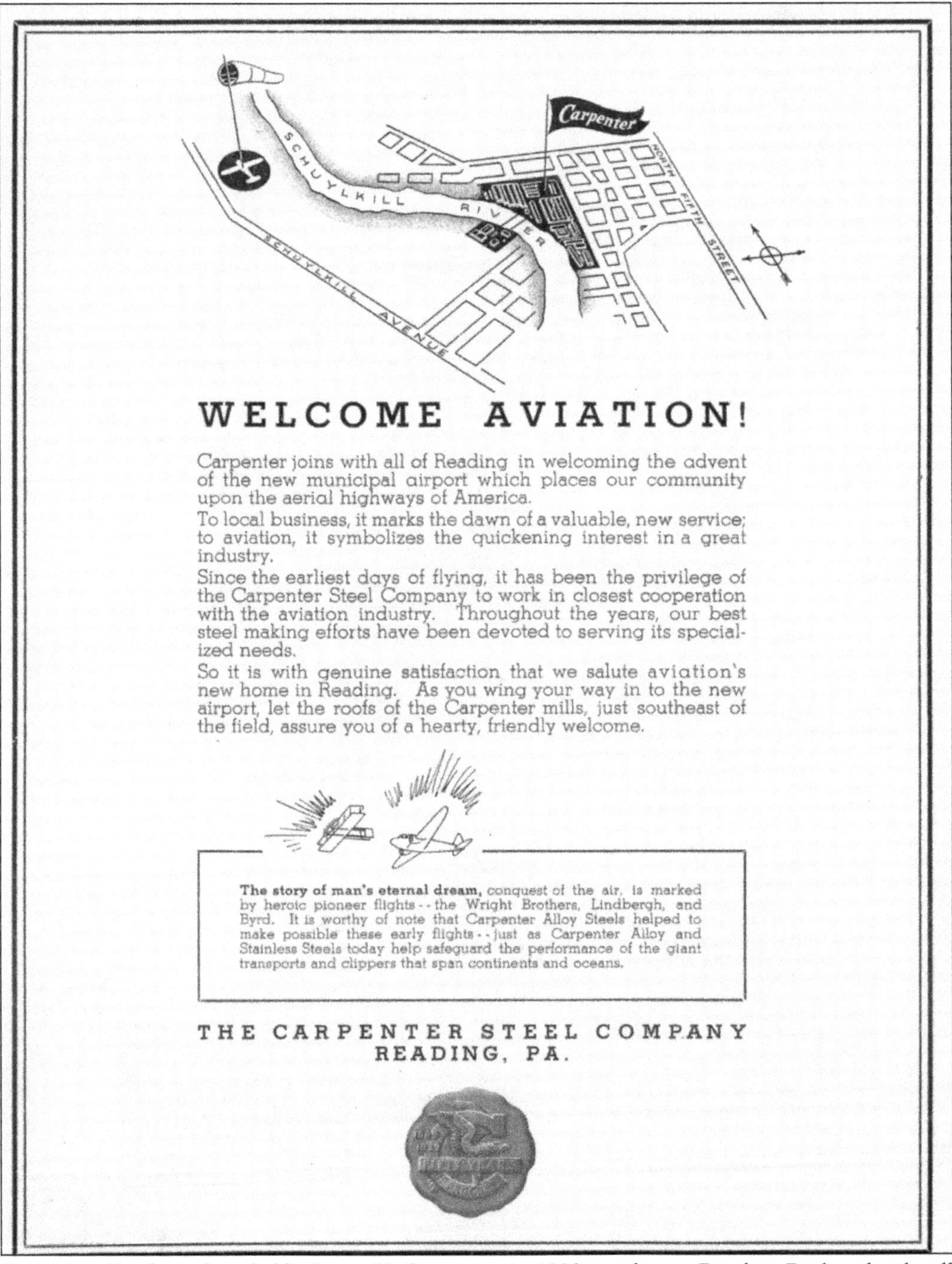

Carpenter Steel was founded by James H. Carpenter in 1889 in a former Reading Railroad rail mill on North Ninth Street. Throughout the years, the company made products for many industries; in particular, armor-piercing shells for the Navy. Its alloy steel was used in the manufacturing of aircraft. (Courtesy of Reading Airport.)

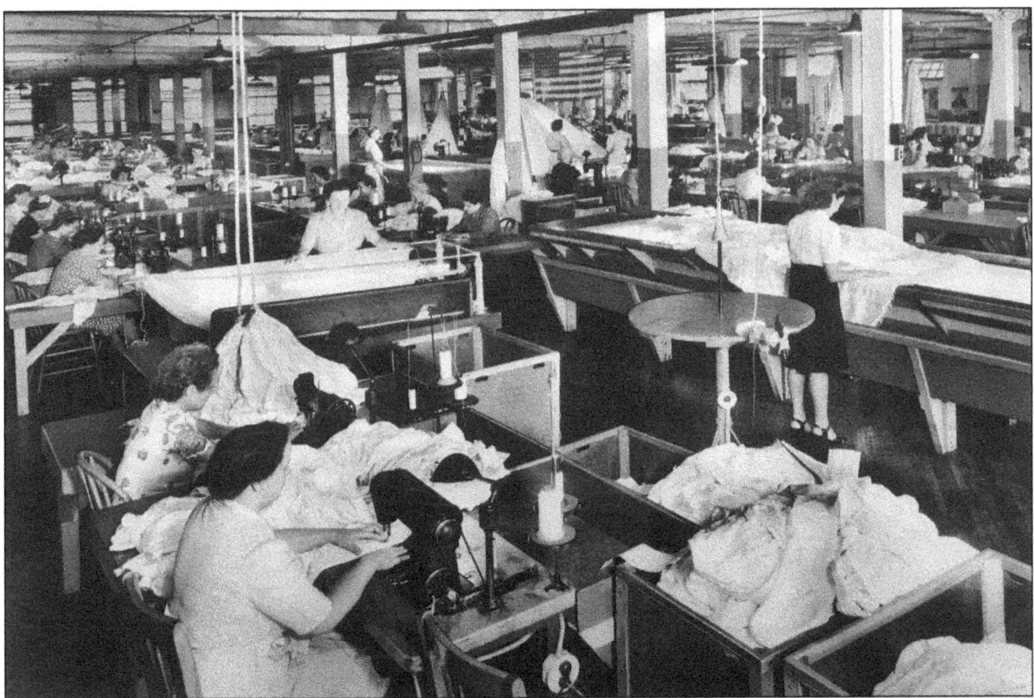

Reading Air Chutes Inc. was founded in 1941 and became the largest manufacturer of parachutes for the military during World War II, earning awards for its bomb, flare, and cargo parachutes. Above is the sewing department, with gore seaming, hemming, and inspection in progress. Below, a woman machine-stitches a parachute. The company eventually switched to producing swimwear and is now known as Dolfin. (Both, courtesy of Berks History Center.)

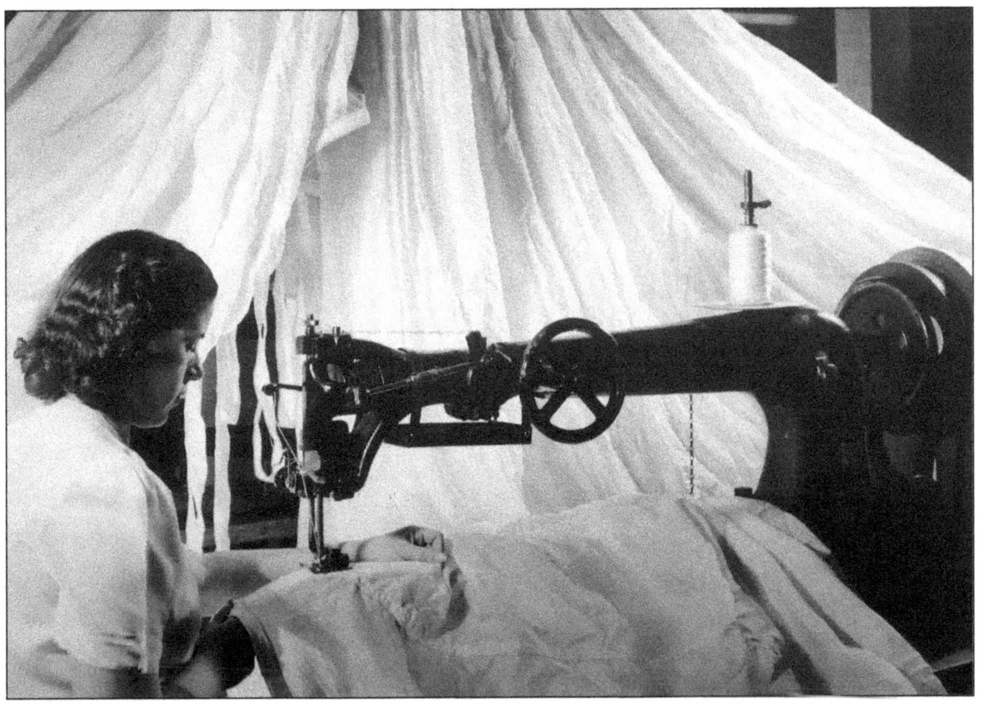

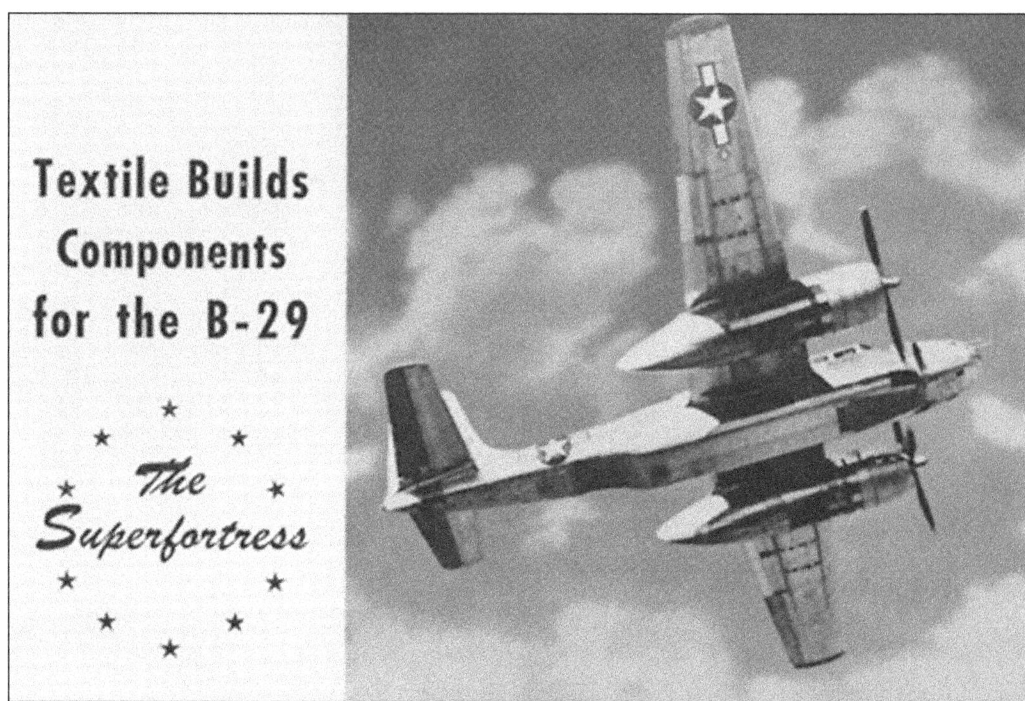

The hosiery industry in Reading and Berks County faced significant challenges during World War II. Japanese silk was previously used for stockings, but war with Japan prevented the United States from obtaining this raw material. Synthetic nylon was introduced shortly before the war, but it also became rationed. During the war years, production of the "Reading" machines at the Textile Machine Works and the Wyomissing Industries foundry ceased entirely. Army-Navy E Award pennants were awarded to the plant for its intensive armament production during the war. (Both, author's collection.)

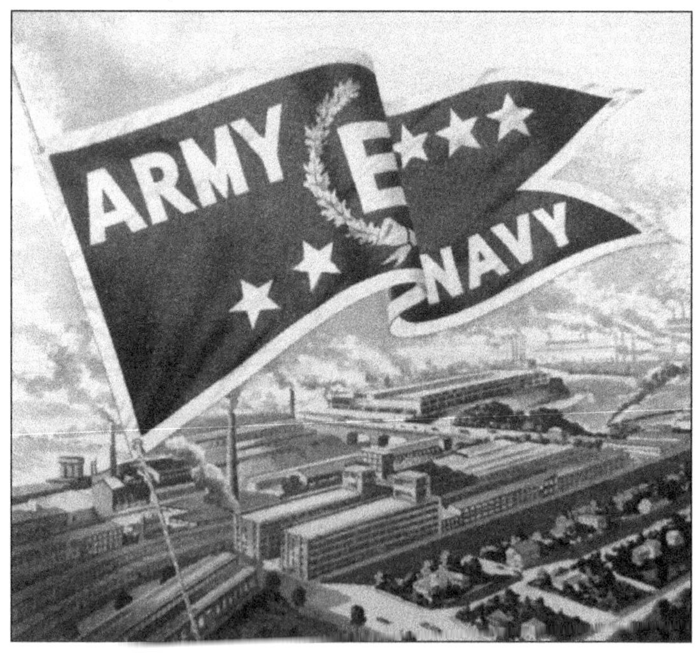

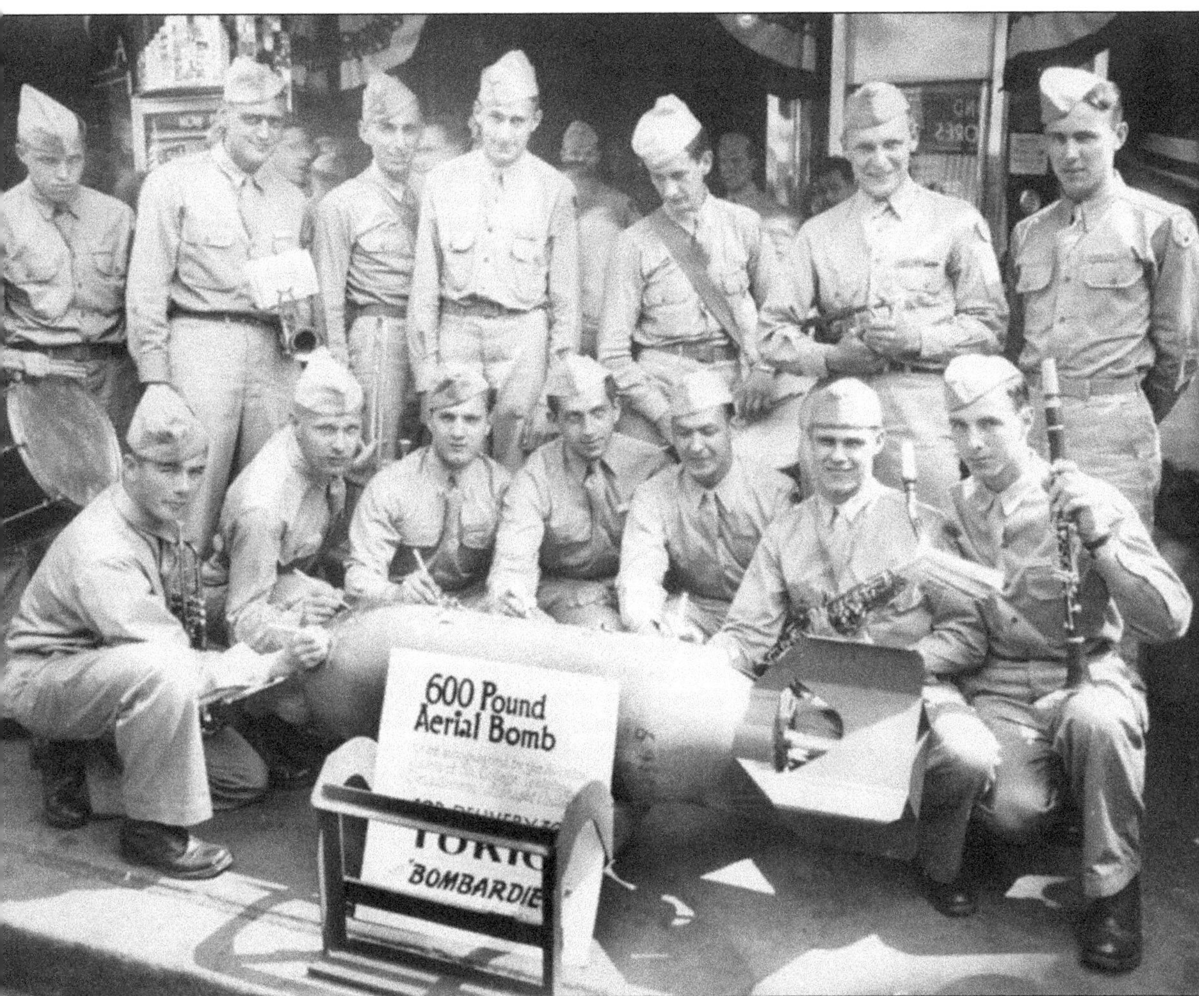

During World War II, Dana Corporation's Parish division converted its operations to aid the war effort. Bombs such as this were one of the products made by the company. The plant expanded five times from 1941 to 1943 to keep up with military orders, including gun mounts and aircraft parts. This photograph, taken during the war, shows the 7th College Training Detachment of Albright College cadets in front of the Astor Theatre, signing a 600-pound bomb "For Delivery to Tokyo." (Courtesy of Albright College.)

www.ingramcontent.com/pod-product-compliance
Lightning Source LLC
LaVergne TN
LVHW060624110826
845147LV00015B/929
* 9 7 8 1 4 6 7 1 0 3 2 2 0 *